TRAVEL WRITE

Select entries from 20 years of the Bradt Travel-writing Competition

TRAVEL WRITE

Select entries from 20 years of the Bradt Travel-writing Competition

EDITED BY CELIA DILLOW

First published in the UK in September 2021 by
Bradt Guides Ltd
31a High Street, Chesham, HP5 1BW, England
www.bradtguides.com

Print edition published in the USA by The Globe Pequot Press Inc,
PO Box 480, Guilford, Connecticut 06437-0480

Cover design by Ian Spick
Front cover image by Adolfo Félix/Unsplash
Layout and typesetting by Ian Spick
Production managed by Sue Cooper, Bradt Guides & Zenith Media

ISBN: 9781784778491

British Library Cataloguing in Publication Data
A catalogue record for this book is available from the British Library

Digital conversion by www.dataworks.co.in
Printed in the UK by Zenith Media

CONTENTS

FOREWORD

1 FIRST WORDS

2 PLACES & SPACES

3 MEET & GREET

4 FEATHERS, FINS & FUR

5 THRILLS & CHILLS

7 GHOSTS & DUST

8 PASSPORTS & PRIZES

FOREWORD

I'm late writing this foreword – with good reason. I thought I'd just speed read a few of the pieces but that wasn't possible. Every story carried a memory of the pleasure of reading it for the first time during the judging process, when all the entries that follow elicited a delighted smile and a muttered 'Oh yes!' Some reduced me to tears, some to laughter, while others had me wide-eyed in horror at a nerve-wracking adventure. All are exceptionally well written, with each word carefully chosen, and skilfully planned to ensure that the specified theme is embraced.

The theme is central to our annual competition. It's hard enough to write really well about travel without any constraint on subject matter. It takes an exceptionally skilled writer to shape their story around a given theme.

The Bradt Travel-writing Competition has a long history. When the company had its twenty-first birthday, back in 1995, I wanted to mark the occasion with a writing competition and was lucky enough to forge a relationship with *BBC Wildlife Magazine* for a few years for travel stories with a natural-history slant. Then, in 2004, our 30th anniversary, we found a new partner in the *Independent on Sunday*, and relaunched the competition with the theme 'We've Come a Long Way' – as indeed we had! And there it stayed until *The Indy* ceased its print edition in 2016. Two constants, throughout the competition's history, have been the terrific prize of a holiday and a commission to write about that trip. There has been a close relationship too with Stanfords in Covent Garden, which always hosted the awards ceremony.

You've only got to read the biography at the end of each piece to appreciate how much winning this competition, or being shortlisted, has helped the career of the writer. But not all the stories in this collection were winners. Some were finalists or particular favourites in the shortlist, but all had something special about them which prompted Celia to include them here.

You are in for a treat!

Hilary Bradt

1
FIRST WORDS

"The competition and the whole experience (at the awards ceremony) was what gave me the confidence to submit my novel to agents. I'll forever be grateful."

Hannah Doyle

The Bradt Travel-writing Competition is a high-profile and well-established event. It always attracts a wide field of entrants from across the world and the prizes are usually spectacular. It formed the structure of my writing year for a long time and when I received 'that' email, from Hilary Bradt herself, saying that my piece had been shortlisted, I was hooked. It was the encouragement and kindness of the team at Bradt which kept me looking out for the launch date each year; noting the deadline; puzzling over the theme.

FOR WRITERS

This spring, as I contacted the authors for this anthology, I realised that many of us shared that experience. Bradt does a generous job of encouraging emerging writers. So, for the first time, all the winning entries are collected together in one place, along with a good selection of others that have caught the judges' eyes. If you want to know how to succeed in the Bradt Travel-writing Competition, this anthology is a must-read.

FOR READERS

We all need armchair adventures in these travel-starved times and this is a delightful collection of quick reads about memorable journeys. Join us as we zigzag across the globe with our writers. Feel their fear, gaze in wonder and laugh out loud. Through these sparkling stories you can reach the whole world: climb above the clouds, cross the oceans or wander in markets and deserts. You will suffer altitude sickness and sunstroke; fall down drains and dodge officials, danger and death. Above all, you will meet some unforgettable characters. As we celebrate our urge to see over the horizon, these stories remind us that we are totally different and so very much alike.

800 WORDS

The word limit in the competition is 800 words. For a writer, an 800-word essay is a great test. It is long enough to develop an idea or two, find an angle and explore a theme. And it's tight enough to be a challenge. You must keep your focus, maintain the tension and make every word count.

For a reader, on the other hand, 800 words is the perfect quick read. It is long enough to get a powerful punch of place or a dose of drama, but short enough to consume with a cup of tea during a quick break.

So join us and jump in. Who knows where your reading will take you?

Celia Dillow

2
PLACES & SPACES

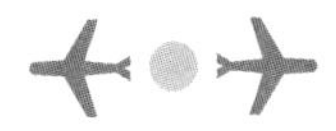

"A bruised and bloodshot evening sky gave way to a morning bandaged in mist."
Moira Ashley

This is writing that drags the reader into the scene and cries, 'this is what I saw and felt and heard.'

India Cuba France USA Egypt Italy Sardinia Norway Iceland Australia Wales Nepal Costa Rica

THE BOATMAN

Louise Heal

Winner 2007 'A Chance Encounter'

The river was at the bottom of a long road. I walked downhill past the Orthodox Cathedral, the High School and the doctors' surgeries. I followed the road all the way to its end, to the forecourt of 'AVG Motors Ltd'. After that there was just the river.

A small wooden shelter stood by the bank and I sat down on the bench. The river flowed slowly from left to right and water lilies grew along the edges. A rope was tied to a palm tree and stretched across the river to another shelter framed by palm trees. Behind the trees were acres of paddy fields. Behind the fields were forests and hills whose tops vanished into grey clouds. A long blue train moved slowly across the fields.

I could hear the faint sounds of car horns in the town, but they were far away. Here was birdsong, running water and the occasional voice. A kingfisher flashed in front of my eyes and landed on a branch. A cow moo-ed in a field and the kingfisher dropped into the water before flying off again. I could not remember the last time I had been somewhere this quiet. It was a perfect Sunday afternoon.

An old man wearing an orange shirt and dark dhoti poled a wooden canoe along the opposite shore to the shelter. A young girl walked out from behind the shelter and climbed aboard. The old man put down his wooden pole and took hold of the rope. He pulled until the boat pointed towards me. Then he pulled hand over hand on the rope and they came across the river. The girl paid him and jumped out. The old man tied his boat up and sat down in the shelter next to me. He nodded in greeting.

I looked at him. He had a white moustache, a white beard with flecks of black and black hair with flecks of white. He sat cross-legged and the veins on his calves stood out. I could also see the veins on the backs of his hands. He was no more than 5' 4" tall and his frame was tiny, with no evidence of body fat. At home, I would have estimated his age at 70, but in India it was harder to tell.

A small house stood upstream on the riverbank opposite, with a tall tree by its front door. A little girl in a white dress stood under the tree next to a large pile of laundry. A woman crouched in the water, scrubbing at a garment, and the soap-suds floated away downstream.

Suddenly the wind came up and it began to rain hard. Gusts of wind swept down the river and blew the raindrops along the surface. Two thunderclaps sounded and the rain fell harder. The drops engulfed the water lilies until their leaves dipped underwater. Clouds descended over the forests and the sky was a uniform steel grey.

The woman took the little girl by the hand and they ran into the house. The pile of laundry remained under the tree. The old man stayed with me in the shelter. A young boy, holding a bag of books over his head, ran down the road and joined us.

Just as suddenly, the rain seemed to have stopped. I could see no drops on the river and the water lilies were still once more. But I looked towards the garage and saw that the rain was still falling, in a mist so fine that the droplets were barely visible.

Then the clouds lifted above the hills and a patch of blue sky appeared. The wind began to die down. The woman and the little girl came out of the house and the woman went back to her laundry. The little girl stared at us across the river.

The old man untied the rope and went across the river with the young boy as passenger. The boy walked away down the riverbank and disappeared from view.

A man and boy walked up the river, chattering as they went. The old man was busy winding a turban on to his head, but waved at them in acknowledgement. When they boarded the boat, he pulled them across the river and came back to sit next to me.

The wind had now died down completely and the river flowed gently again, with the palm trees clearly reflected in the water. The kingfisher flew back and sat on the rope, casting a royal blue glow on the water beneath him.

The old man sat silently, waiting for his next passenger. I got up to leave and said goodbye to him.

He smiled and raised his hand in farewell.

Louise Heal *has worked in the City of London for over twenty years. During that time, she has travelled extensively in India, particularly Kerala. She recently graduated with Distinction from the MA in Creative Non-Fiction at City, University of London, and is working on her first book about Anjengo Fort in Kerala. She is married and lives in southwest London.*

THE PERFECTION OF IMPROVISATION

Kate Megeary

Winner 2008 'The Heart of the City'

A small brown dog wearing a faded pink T-shirt jogged down a dirt street. I decided to follow him. He seemed as good a guide as any. He was in no rush, stopping to sniff the flip-flopped feet of the girls who sat gossiping on doorsteps, rocking babies, their tight, skimpy vests revealing cleavage you could lose an arm in.

The dog took me down the narrow back streets of Old Havana, where faded pastel paint peeled from the façades of elderly buildings. White sheets and blue shirts were strung up to dry over the twisted stumps of wrought-iron balconies; women leant over crumbling carved stone balustrades and shouted to their children in the street. These once pristine and exclusive colonial mansions quietly decay while life inside them thrives. Here lies the beauty of Habana Vieja, evident in her decline.

My guide stopped, his ears pricked, as he spotted a man in torn denim shorts with a thick gold chain around his neck. The man wiped sweat from his armpits with a handkerchief and shouted up at a window high above the street. A woman with curlers in her hair leant out of the window and lowered a wicker basket on a piece of rope. The man took his pizza out of the basket, replaced it with a banknote and the basket was raised again.

A gang of kids with skinny legs and grubby T-shirts had set up a baseball pitch at a crossroads, each pavement corner representing a base. I stopped to let them pitch. 'Hey beautiful lady,' called a small boy, smiling mischievously at me as he threw a small coconut. Another boy hit the makeshift ball expertly with a stick. The dog caught the ball in its mouth. The children shouted. The dog ran. I was guideless once again.

The uneven dirt streets gave way abruptly to newly laid cobbles and opened out on to Plaza de la Catedral. The limestone cathedral was weathered by centuries of hurricanes. Fossilised sea creatures were embedded in its walls, as though the building itself had risen, fully formed, from the sea. Waistcoated waiters served overpriced mojitos to tourists wearing Che Guevara T-shirts, expensive cameras slung around their necks. A brass band played 'Guantanamera'. Ancient black ladies with their life stories etched on their faces wore gaudy satin flamenco dresses and flowers in their hair. They posed for the tourists, huge Cuban cigars dangling, unlit, from their lips.

I bought an ice cream from the ground floor window of someone's house and rested on a bench. A good-looking young man sat down next to me and asked where I was from. We talked, he in broken English, I in tentative Spanish. Suddenly, he stood up. I turned and saw a policeman standing silently nearby, arms across his chest, staring at the man as he walked away.

The Caribbean sun began to lose its heat. I sat outside on the terrace of a hotel bar. The staff looked bored and tired. The bar was empty except for Yamila. She was beautiful, with waist-length hair and daring eyes. She told me she was learning English. That she wanted to travel. An overweight tourist with grey hair and a pink silk shirt sat down at the bar and ordered a cocktail. Yamila excused herself and went over to him.

I gazed at the ferries crossing Havana Bay while the sea turned orange, then grey. When I could no longer make out the white star of the Cuban flag that fluttered above the ferry terminal, Yamila reappeared with flushed cheeks and tousled hair. 'Vamos,' she said, offering me her hand.

She took me to the Malecón. Sweeping around the northern edge of the city, the Malecón's protective wall shelters Havana from the sea. Locals sat hip to hip along the wall, playing music, fishing, dancing, swapping stories, sharing worries, selling peanuts, looking north across the sea.

Yamila's friends were waiting. We drank rum and watched the lights of Havana rippling in the stinking black water below the wall. 'You must be hungry,' said Yamila.

A boy was sitting on the wall next to us, his pole in the water, two unidentifiable fish by his side. Yamila gave the boy a peso and walked away with the fish. She returned ten minutes later carrying freshly fried fish and some rice on the lid of a cardboard box.

She took her ID card from her purse and showed it to me, proudly pointing out her photo, her name, her date of birth, explaining that Cubans must carry this card with them at all times. Laughing, she cut a slice of fish with the edge of the card and scooped it up along with some rice. Nodding encouragement and grinning, she offered it to me.

***Kate Megeary** was a 30-year-old footloose and fancy-free adventure tour leader when she wrote this piece. She now lives in a hobbit house in the wilds of Cornwall and is a 45-year-old mother whose life's work is empowering women to restore honour and connection to their sexual and reproductive anatomy and sexuality. You can find her at ⊗ sacredyonicornwall.com*

A GOATLY ENCOUNTER

Margaret Histed

Longlisted 2012 'A Close Encounter'

The goat turned its head and looked at me. It was small for a goat, black, with dainty little hoofs. Its eyes, however, were like all goats' eyes – yellow and evil. It was perched on a small wooden trolley in the middle of the busy marketplace, tethered to a pin with a length of chain.

Intrigued, I stopped. Around me, fruits and vegetables of every kind imaginable spread out in a vast patchwork skirt of colours. Meat was displayed in all its forms: escalopes, filets and gigots; paupiettes, fritons and terrines; andouillettes, fricandeaux and boudins. Fish lay on beds of crushed ice; bees had contributed honey, royal jelly, propolis and pollen to the feast; the sultry odour of spices hung in the air. To one side of me, a tray of upturned lettuces shook out their frilly petticoats and a row of Kilner jars crammed with thick yellow 'graisse de canard' responded with boozy winks. Above me, buttons marched determinedly down the front of no-nonsense print overalls.

The goat was the only example of livestock to be seen in the marketplace that Saturday morning, and I wondered why. Was it for sale, to be butchered and made into goat côtelettes, rillettes or saucissons? I hoped not. In spite of the evil yellow eyes it was a dear little thing. Was it a walking promotion for the sale of goat's milk and cheese? The trolley held nothing apart from the goat with its chain and a bowl of bright green leaves, which it was chewing enthusiastically. And a large quantity of goat droppings, whose production was closely observed by two wide-eyed small boys.

The man standing next to the goat was no doubt the owner. I would ask him about it. He looked just as out of place as the goat, being decidedly overdressed for a hot day in the south of France in early September. His heavy chocolate-brown woollen cardigan was zipped up to the neck and his denim-clad legs ended in a pair of sturdy boots. I approached him and asked, in my halting French, if the goat was for sale. No, it wasn't. It was a female of the dwarf variety and lived in the rescue centre he ran for abandoned and ill-treated animals, along with two other goats and sundry cats, dogs and rabbits. The goats had once been the property of an elderly lady who was no longer able to care for them. If I would like to buy (here he opened a cupboard underneath the trolley and produced his wares) a packet or tin of sweets (contents: sugar, glucose, honey and balsamic pine flavouring), this would help cover the costs of food and veterinary treatment for the animals. He could do me a special deal if I bought both the packet and the tin. And the goat was called Brigitte.

It was impossible to say no. I looked at the happily munching Brigitte, who obviously had a large appetite, then back at the perspiring Frenchman gazing at me with the soulful eyes of a modern-day St Francis, and handed over my euros. The sweets held no appeal for me but the tin was unusual and would look good among the others lined up on my kitchen shelves at home.

Next stop was a bakery stall, where I bought a 'Jésuite' – a flaky-pastry confection filled with almond-flavoured crème pâtissière. I found a sunny bench and got down to the serious business of eating. Mmm! Four euros, but worth every cent.

Four euros... Something clicked in my head. How much had I given the goat man? A quick calculation and I realised the dreadful truth – it was easily enough to keep Brigitte and her pals in Bacardi and Coke and fuchsia-pink nail polish for the next month.

What on earth had come over me? He must have sussed me out the minute he saw me. A lone female tourist – and an Englishwoman at that, guaranteed to be as soppy about small furry animals as the rest of her race – would be a walkover. No doubt Brigitte had gone without breakfast that morning so that she would appear all the more endearingly hungry by the time I saw her. As for the two small boys so interested in her bowel movements, they were probably St Francis's sons, being trained for a life of crime. Come to think of it, they had the same hypnotic brown eyes…

A Londoner now living in Bristol, ***Margaret Histed*** *is a freelance publishers' editor and writes in her spare time. Her work has appeared in book, magazine and newspaper form, her one-act plays have been staged, and her stories have been broadcast on BBC Radio Bristol. 'A Goatly Encounter' was written during a creative-writing holiday in the south of France.*

BREATH-TAKING BRYCE

Moira Ashley

Finalist 2019 'Out of the Blue'

'You gonna have slaw with the pulled pork? Awesome! Where're you from?'

Wales – 'Is that in Yurup or Scotland?' – was, it transpired, just as awesome as my choice of dressing.

Moments later the waitress was repeating the exclamation at the next table (two cheeseburgers, Seattle). I soon discovered, however, that Kimberlee's high-octane enthusiasm did not extend to her surroundings. This was Panguitch, Utah, less than twenty-five miles from the geological wonder of Bryce Canyon. Kimberlee, a local girl, had never been tempted to visit and could only advise, 'I guess you'll find a whole buncha rocks.'

I had come to Utah to hike the National Parks; the weather, however, had gone rogue. TV reports warned that the violent storms, which had already claimed several lives in Colorado and had now moved south and west, showed little sign of abating. Zion and Bryce had been closed for three days, their normally dusty tracks now swirling rivers of rust. I huddled in the steamy diner, hail and thunder warring overhead. Leaning my head against the rain-streaked window I sighed gustily, not for the first time that day.

A bruised and bloodshot evening sky gave way to a morning bandaged in mist. Good news, though: Bryce had reopened. I splashed along through puddles of burnt umber, windscreen wipers impatiently trying to rub a hole in the clinging mizzle. I might as well have been in a Lincolnshire cabbage field than within yodelling distance of the iconic wind-sculpted hoodoos. According to Native American legend, these totem-pole-like formations were originally a local tribe, turned to stone by the coyote god for their mistreatment of the land. Inching forward in blind hope, I finally reached the park entrance.

First stop was for a coffee at Ruby's Inn where the smiling staff were keen to feed me titbits about the park's history.

'Bryce is named after Ebenezer Bryce, a Mormon pioneer who came to settle in the area in the 1850s,' explained Chad as he brought my drink. 'He and his wife built a ranch in the valley and grazed cattle.'

'Word is, old Ebenezer didn't find nothing to get excited about when he first set eyes on the hoodoos,' the barista chimed in. 'Just said it would be a helluva place to lose a cow. Folks was more practical in them days. Saw rocks and such as getting in the way of farming.'

Fuelled with caffeine and renewed determination, I headed out. On my trek across the plateau I could see that the mist was thinning; the pale tips of greenleaf manzanita shrubs poked defiantly through its weakening tendrils. Soon it seemed I could almost hear the last remaining shreds sigh as they dissolved in the rapidly warming air. And then, as I approached the rim, a shaft of sunlight, biblical in its intensity and suddenness burst on the horizon and an azure hue spread across the sky revealing a vista of dazzling beauty. Mile upon mesmerising mile of magically fashioned rocks stretched out in a glowing spectrum: from ochre through cinnamon, coral and apricot to dried putty.

My first, shimmery impression was of a petrified forest, autumnal foliage blazing. When I zoomed in on detail, the imagery became increasingly fanciful: over there, I decided, was the Great Wall of China flanked by a platoon of terracotta warriors. Here, between that trio of gossiping, white-wimpled nuns and Gothic cathedral windows, were giant strawberry ice lollies, topped with whipped cream. Surely that rose-pink hoodoo in the middle distance bore more than a passing resemblance to Queen Victoria in unamused profile? Eventually though, the metaphors petered out as I gazed at the panorama before me. I could find no better word than Kimberlee's: it was simply awesome.

Moira Ashley *(🌐 moiraashley.co.uk) is a retired lecturer who has enjoyed wide-ranging success in travel-writing competitions including SAGA, Rough Guides and the* Daily Telegraph *(the fruits of which have funded further travel!). She has also been published in several poetry anthologies. In 2019 she won first prize in the 'Short Memoir' category at the Charroux International Literary Festival. The year 2020 confirmed her belief that travel writers can find rich sources of inspiration on their doorstep, as some of her recent writing illustrates.*

A WALK IN THE DESERT

Jean Ashbury

Longlisted 2014 'Meeting the Challenge'

A dust devil rose from the desert floor like a phantom. It approached with the speed of a Cairo taxi and found me in its way.

That morning began with silence so heavy and dead I yelled to prove I was still alive. Fog hung over the dunes. Winter chill nipped my nose and stung my fingertips. Around me, slithery trails and birdlike tracks told me desert creatures had checked me out while I slept.

I emerged from my sleeping bag dressed in a week-old stink of clothes. I shook my boots and sent scorpions scuttling, tails curled and sting ready, before shoeing my blistered feet. Wood smoke drew me to our dawn campfire. I sat beside it like a half-wrapped mummy thawing back to life.

Over gallons of sugary tea, Mahjdi and his desert 'commandos' discussed the day. Our walk to the White Desert near the oasis of Farafra would be long, exposed, and incinerator hot so we must fill up like camels before heading out.

'You OK?' Mahjdi looked at me with concern.

I'd been trouble all week, passing out with dehydration, and whining about sunburnt lips, heat rash and bruises from camels. All week his espresso-coloured eyes had asked, Why did you come?

To unclog my brain was the silent answer. I was convinced that ten days on the Great Desert Circuit from Luxor to Cairo, stopping off at four oases to walk in the desert and sleep under the stars would do the trick. Just a tough holiday, I thought.

But in Luxor, the guides' briefing to our small band of trekkers rang alarms – 'Drink, pee, drink. Cover up or you'll fry. Don't wander off anywhere. This is the Western Desert, not the Costa del Sol.'

When they finished, I realised I was on an 'expedition'. Days were spent walking between dunes where footsteps left no trace. Where we went looked like where we'd come from, and every dune appeared the same. My skin flaked like old bark, and my constant mirage was cold beer with condensation running down the glass.

We set out that morning in icy light. Mahjdi was on straggler duty with me and my camel, a beast the colour of bleached rope with burnt tufts. Though slowest in the pack, her lolloping gait was still too fast for me. I kept in step with the spindly calf, tiptoeing behind as if in high-heeled shoes.

Midday heat shooting off the thermometer caught us strung out on the razor edge of a crescent dune. One sleek side sloped to the ground with glassy crystals glinting amid golden grains of sand. The other rippled with waves as the seashore does when the tide flows in. Some poetry sprang to mind, but this was no time for words. I guzzled tepid water and taking my boots off soothed my feet in sugary sand cool as the sea.

Shimmering heat haze guided us to the White Desert. Light, diamond hard, played on giant white monoliths, mushroom sculptures and creamy boulders strewn on a bed of sand the colour of cooked pastry. It could have been a meringue cake made by gods. I sat beside my hobbled camel, eating falafels (chickpea fritters) and listening to the wind rush by. That same wind had sculpted these chalky shapes when an ancient sea floor dried up. Beside me, seashells from millennia ago stuck out from a limestone wall. I picked up a fossil of polished black stone with spikes like a mace, my souvenir of time past.

And then they came – a corps de ballet of dust spouts and its principal dancer, desert gatekeepers demanding an entry fee.

'Djinns. Spirits of people dead in the desert. They will pass Insha'Allah,' said Mahjdi.

Tucked into the side of my camel, I remembered the houses and telegraph poles I'd seen sticking out under marching dunes, and hoped Mahjdi's Allah would be merciful.

Whooshing threats to bury me, the djinns rained over for a lifetime with a blizzard of grit and sand. When they were spent, my eyes, ears, nose and every part of me was filled with sand. Not the soft mush of sandcastles by the beach, but sharp particles of rock that turned me into human sandpaper scraping and grating between teeth, between legs and under armpits.

I was alive, though, and I thought I heard the universe click. I'd beaten the desert.

Later, at peace in my musty sleeping bag, I recalled legends I'd read about the Western Desert, of lost armies and mythical creatures. I watched the sky turn pink then orange and become black and clustered with stars. I felt every tingle in the air and just before I fell asleep, I added my legend to the list – I was here.

Jean Ashbury *is from Trinidad in the Caribbean, but lives in London, and has her heart in both places. She is a teacher, writer and traveller with writing in anthologies for Jerwood/Arvon, the London Short Story Prize, and others published by Bradt Guides, She Voices Writers' Group, Writers Abroad, and Kingston University.*

AT THE RIALTO

Liz Sillars

Finalist 2008 'The Heart of the City'

Just for a few days, I want to pretend that I live here. I take my string bag and enough money and head to the market like the other Italian housewives who nod to their acquaintances in the street. I need to be there early to get my pick of the freshest vegetables and the choicest fish. And I don't want to be mistaken for a digital-camera-wielding tourist either, because I am on my way to the Rialto.

Venice was a great city-state whose wealth was founded on commerce between east and west. While St Mark's was about pomp and power and politics, the Rialto was about trade. The doge has gone and the Council of Ten is disbanded but every weekday morning you can go to the market and hand over coins in return for goods from the Orient.

When Shylock asks, in *The Merchant of Venice*, 'What news on the Rialto?' he's asking for a business update of the world markets. The Rialto was the trading floor and exchange of the empire with the Banco del Giro housed in a building near the bridge.

Today, the high finance may have been replaced by low tourist tat but it is still the mercantile centre. If we don't buy the fake replica football shirts or glass gondolas then the traders will supply us with something else that we need: perhaps a chic leather handbag with a logo a little like Prada's. But I am here for the serious business of food.

Offloaded before dawn at the evocatively named Fondamenta dell'Olio from a fleet of jostling barges, the goods are now ready for my inspection beneath the stone arches of the Erberia. I am helped in my role-playing by the fact that most items are labelled, albeit in the Venetian dialect, but a little pointing and smiling goes a long way for a non-Italian speaker. The artichoke man, almost a caricature of his produce in a green jumper offset with a lot of chunky gold jewellery, is offering purplish Romani as well as the green Bari types. With universal gestures, he recommends the tiny Castravre, but I can only wonder at what their name might mean.

My next stop is at the flower stall where a matronly woman sits like a sunflower in a vase surrounded by colour. My halting efforts in Italian to ask for a bunch of ink-blue irises are rewarded when she sneaks in some extra blooms and ties them with an exuberant bow.

At a brightly lit stall in the fish market, clams from the lagoon are sold in a net bag but some have escaped and every now and then one jumps in a futile bid for freedom. The eels too are unhappy at their confinement and attempt to slither out of their trays. The cuttlefish, thankfully, have already expired or they could be a menace with their two-foot-long tentacles. Overseeing them all, the fishmonger sings out the praises of his merchandise while keeping them husbanded within the confines of his stall. He uses his green watering can to keep the fish looking fresh and tempting.

Round the corner is a traditional butcher's shop, with a white-aproned proprietor behind the counter. He greets his customers, guides them in their selections and wraps their purchases with an operatic air of bonhomie. The lettering above the door spells out Marcelleria Equina. Most of the trays on display in the window contain variations on the theme of mince but there are also joints of meat including *fianchietto di puledro*. I have to look that up surreptitiously in my dictionary and when I learn that *puledro* means 'foal', I decide I don't need to research any further.

Almost next door is the cheese shop. Ah, this is better! The enticing interior has Pirelli tyres of Grana Padano, soft white curds in china bowls and a fragrance that makes me want to rush out and eat pizza. Never mind that most of the produce is from Emilia-Romagna; it's unlikely that most visitors to Venice will be progressing on a grand tour so why not pick up the parmesan from a proper cheesemonger? They won't be taking home much liver, sour onion or other Venetian specialities in any case.

The market is winding down now; skinny cats scavenge for fishy offcuts, cabbage leaves are brushed into the gutters and metal tables hosed down with workaday efficiency. I'm also getting a bit tired, clutching my bouquet like a diva. I think it may be time to call it a day and head home. But first I must do what the Venetians do: go to the supermarket and do my shopping.

This skilfully told tale was a finalist in the Bradt Travel-writing Competition in 2008. We have since lost contact with the author.

OLBIA

John Carter
Finalist 2005 'If Only I'd Known'

The ancient engine dragged its carriages along the winding single track from Sassari, reached Olbia by the skin of its teeth and stopped, exhausted, in the station.

According to the tourist brochure I had been reading during the journey, Olbia, that sizzling July day, was either: 'contemplating its bustly harbour,' or 'looking with nervousness towards the banditted hills.' Either way, it wasn't expecting me.

Despite my letter of introduction from the Italian authorities in London, the young man in the tourist office regretted he could be of little help. No hotel rooms were available. No warning of my arrival had been received.

There was a long pause, and silence save for the slow beating of the ceiling fan and the cries of some boys playing in the dusty alley. Impasse.

We tried to find a solution. The young man – I think his name was Antonio – suggested we might, first, find refreshment. So we left the office and walked towards the Corso Umberto.

This was Olbia's main street – actually, the only street that went anywhere. Others performed gyrations around it, dodging round houses and under washing lines, through hen-strewn courtyards and back to the Corso again.

We crossed the square to the café of Signorina Sophia, and, over coffee, solved the accommodation problem. A room in a private house could be obtained. Was that acceptable? It was, so the boys from the alley who had trailed us to the café in hopes of a cream cake were sent back to the office to get my bag and take it to the house.

It was a gem. In one of those meandering back streets, it had a dark, marble-floored courtyard set about with tubs of blue hydrangeas. A widow owned it, and her daughter was also recently bereaved. Antonio asked if I minded being in a house with two widows – he seemed to think it was a bad omen. I said I did not mind, and was shown my room.

Payment was arranged with discreet dignity. Then I expressed my sadness at the recent death of the son-in-law of the Signora, the husband of the young Signora, and this was passed on with due formality by Antonio, who then left, promising to meet later in the café so we might discuss important affairs.

So began a dozen of the finest days I can remember from those long-ago times when I travelled alone, but was rarely lonely. In a town I did not know, among people whose language I did not then speak, I became less of a stranger.

Each morning, the clock in Signorina Sophia's café was my first indication of time, for my watch was broken and the house of the two widows seemed to possess no clocks at all. I'd ask for coffee, select a plate of buns and eat them under Signorina Sophia's benevolent supervision.

She laughed when I read the newspaper, but was pleased when I said '*grazie*' and '*prego*' and mastered a few other simple words. She did not know I had a phrase book in my suitcase, studying it in the white-walled bedroom of the cool house. Lying on the creaking double bed with its black iron frame and wobbly brass globes. Watched over by a soulful Jesus in a wooden frame.

With Antonio and his friends, I went in a battered blue Fiat up and down the coastline to their favourite bathing beaches. But one day we encountered two men, bent beneath the sun with dust streaking their bare and sweating backs as they built a wall across the track leading down to the shore.

There was an argument… a waving of arms… a grudging explanation… a spitting of curses. Some rich foreigner, 'from Manchester in England', I was told, had bought the land. The beach was his. Private now. Keep off. That evening in Signorina Sophia's café, we drank a lot of cold beer and cursed such foreigners. Antonio explained that much land was being purchased along that part of the Sardinian coast but nobody seemed to know why, or to care much.

Next morning, when I left, he and the others came to the house, shook hands gravely and said important things I did not understand. The two widowed Signoras stood beside the hydrangeas and I bade them farewell. The solemnity of the moment was broken by the arrival of Signorina Sophia who embraced me to her ample bosom and presented me with a bag of cream cakes for the journey. Then she cried.

As the little train twisted its weary way back through the 'banditted' hills, I shed a tear as well.

Had I known that simple shoreline would, quite soon, be transformed into the Costa Smeralda, I would probably have wept until my journey's end.

***John Carter**, at 86, is retired, though he has a couple of writing projects on hand, one of which is another collection of stories to follow* Gullible's Travels, *which Bradt published in 2016. He is Patron of the Silver Travel Advisor, contributing a regular article to its website, and a Patron of the Family Holiday Association. In his previous life he was Travel Correspondent of* The Times *(its first), Travel Editor of* Good Housekeeping, *and presenter/reporter on BBC television's* Holiday *programme and Thames Television's* Wish You Were Here…?

RUBY

Eric Baldauf

Longlisted 2013 'A Narrow Escape'

Las Vegas glimmered from miles away like a desert jewel. As I drove down 'The Strip' for the first time, I had to blink my eyes over and over, adjusting to the bling of neon, to the lustre of waxed limousine and the shock of the whacked-out, ostentatious architecture. But it wasn't just my sense of sight that felt overloaded. The city sounded like onomatopoeia on steroids – the whistling, honking, screeching noises of a town hyped up on the adrenaline of chance. And once I entered the casinos, I was surrounded by the clink of coins dropping from slot machines, the whizz of the roulette wheel, the tinkling of crystal, the whoops of the victorious and the howls of the disappointed.

I was tucking into my one-dollar sirloin when I spied a flash of red entering the casino restaurant. She wore crimson patent-leather pumps and a burgundy chemise; a chunky scarlet cross hung in the valley of her generous cleavage. A cluster of delicate vermilion bangles and a pair of diamond-and-ruby chandelier earrings highlighted her shocking ensemble. Middle-aged but very well-preserved, she'd dyed her long hair silver and tied it in a tight, classy chignon bun.

She marched up to my table, sat down and stuck out her hand. 'Hi, I'm Ruby and I want to play with you.' I was mesmerised. Her pale-blue eyes were full of challenge and her heavily painted mulberry lips were mouth-watering. I practically choked on my steak.

'You new in town?' she asked, though it must have been obvious. I nodded.

'Ever play high-stakes poker?' I shook my head.

'Doesn't matter, I like your face.' She stared intently. 'I assume you know the rules.' Before I could get a word in she continued, 'We'll get you all pimped-out and see what we can do.' I had no idea what she had planned but I knew I'd already been hooked and reeled in like a guppy on a shark-weighted line.

Ruby came back thirty minutes later with some new Levis, a rhinestone-studded western shirt, a pair of aviator Ray-Bans and a broad-rimmed Stetson. 'You oughta go get changed. I have two high-stakes seats booked at the Bellagio and we've got to discuss strategy before we get started.' I had no idea what she was talking about but I was more than happy to go along for the ride. 'Hey cowboy,' she added as I walked away, 'you might want to shower – get rid of that road stink before you put on your new gear.' She smiled and winked.

Her rules were simple. She put up the money. I would get ten per cent of the profit. But I had to stay in the game for a minimum of two hours. If she touched her cross I dropped out of the hand. If she peeked at her cards twice, I stayed in. There were a few other signals. That was it. Lesson over.

We drove to the Bellagio in her red Chevy pick-up. She pulled up a few streets away, put her hand on my thigh, leaned over and pulled my head around with a tug on my ear. 'Remember, we don't know each other. We meet up here an hour after the game's finished.' She handed me a huge wad of crisp one-hundred-dollar bills and kissed me on the lips. 'Good luck and a word of advice, try to lay off the free cocktails.' As I got out of the truck she added, 'And don't fuck up.'

The poker was intense. We both went up quickly, don't ask me how. I kept my Ray-Bans on and tried desperately not to smile. My pile of chips increased slowly, but Ruby's grew into towers. Around

4am the mood suddenly turned nasty. One of the big losers, a drunken attorney from Philadelphia, suddenly hurled himself out of his chair and yelled, 'Something stinks around here.' My heart almost leapt out of my chest. All the players stared at each other accusingly. I felt like a marked stooge in a trite gangster movie. Then Ruby smiled innocently and said, 'Yeah mister, losing always stinks.'

The game finished quickly after that. Ruby and I collected our winnings. When we met up at the rendezvous Ruby gave me deep, sensuous kiss. 'We make a great team.' The cliché rolled off her lips. 'Do you want to go double or nothing?'

I've never been so tempted. Instead, I tucked my huge wodge of bills down her décolletage, leapt out of the pick-up and quickly high-tailed it back into the desert, certain my beginner's luck had run its course.

That one game was the beginning and end of my professional poker playing career. I've never had the urge to return to Vegas. I've never played poker again.

***Eric Baldauf** was born in Texas but settled in London more than twenty-five years ago. He met his English wife in Zanzibar and their mutual love of travel has taken them on many a far-flung adventure.*

NORTHERN LIGHTS

Kenneth Steven

Longlisted 2019 'Out of the Blue'

For a year I lived in Arctic Norway, at the very part where the country is at its thinnest. From the college where I taught I could just see to the west the beginnings of the sea; if I looked east I caught a first glimpse of the Swedish valleys. And between the two ran the main road north, and beside that the railway line. At night I used to watch the yellow caterpillar of the train thundering north and west to the final stations on the line. When I came back, people asked if it hadn't been an ordeal suffering the winter. The paradox is that the worst winter I've ever known is the one *after* I returned to Britain, because the cold here in December is a wet and shivering rawness that's made all the worse by the gnawing of the wind. The Scots have bottled that whole description with their single word *dreich*. The best winter I've ever known was that one, north of the Arctic Circle: an encapsulation of all the childhood dreams of northerness – pine forest and deep mountain and frost-covered snow. And the Northern Lights.

The last thing I used to do at night was to go on to the balcony for a cigarette. It was winter and thirty below: I went out like an extra from *Dr Zhivago*. I will never forget the quiet at ten o'clock: you could have heard a pin drop in Moscow. It was somehow quieter than silence itself: the snow six feet deep and covered with frost crystals. It was what the Norwegians call 'the time of darkness': there was no light whatsoever, just days and days of moon and star darkness. Yet it was a bright darkness; it was a beautiful darkness.

And I went out at ten o'clock every night for the rising of the Northern Lights. I could set my watch by them: when others were switching on their televisions for the news, I went out on to the balcony to await the start of their dancing. For I never knew what colour they might be: that was what held me in suspense every night. Often it was blue and green; sometimes there were ghostly white breaths and flickers, only very occasionally did I witness them vivid red.

But always they were there at ten o'clock in the silence that was bigger than silence, rising above the pine trees and the sharp white edges of the mountains. I thought of their leaping as like that of strange and ancient ghosts; I fabled that once upon a time in the very north of the world there were beautiful horses that galloped and leapt, but something terrible had befallen them and they had died and these were their spirits.

I wasn't alone under the Northern Lights. Other people came out on to their balconies here and there; they waved handkerchiefs at the fire in the sky and I heard them whistling to them. For the Northern Norwegians believed that these things would make the Lights shine brighter still.

What I found eeriest of all was the silence of them. You stand under these great risings and fallings of light and they are soundless. It was so quiet I could hear my own breathing; all I could scent was the pine smoke from chimneys around me, rising in blue pillars into the night. I watched until the cold overcame me and the Lights themselves began to fade, came inside with ice on my mouth and face, not quite able to believe I was really there. It was a childhood dream, and it had happened.

Kenneth Steven *is a full-time writer, journalist and broadcaster. At one time he lived in Arctic Norway where he saw the Northern Lights every evening. He has written a book entitled* Beneath the Ice *about the Sami people of Arctic Scandinavia. For more about all his work, see* 🖰 *kennethsteven.co.uk.*

LAND OF FOG AND TARMAC

Jane Gulliford Lowes

Highly Commended 2016 'A Brief Encounter'

'I'm taking your father out for the day for his birthday. Would you like to come?'

How nice. A leisurely jaunt up the Northumberland coast to windswept Bamburgh Castle? Or perhaps a drive through Weardale, over the Pennines to Westmoreland and on into the Lake District? Maybe a meander along the back roads of the North Yorkshire moors to the ruins of Rievaulx Abbey and a spot of lunch in Helmsley?

'Of course I'd like to come, where are we going?'

'Iceland.'

'The frozen food shop? That's not much of a birthday trip.'

'Not the shop, the country. You know, land of fire, ice and cod wars, all that jazz.'

'Mother you can't go to Iceland for the day. It's almost the Arctic.'

'I can, and we are.'

And so it was that I found myself boarding a small passenger jet at Teesside Airport, one bitterly cold morning in mid-January, with my parents in tow. I envisaged snow-covered plains, windswept ponies, dancing aurora, rosy-cheeked blondes, steaming geysers, grass-roofed cottages, volcanic rumbles, that sort of thing. Mother was beside herself with excitement, Father looked mildly horrified, having only ever holidayed in the Neapolitan Riviera or the Cotswolds. In June. Every June.

'Ooh land of the midnight sun!' exclaimed Mother as we took our seats.

'It's January, Mother. And that's Norway.'

The beautiful Scandic-cool airport at Keflavík was deserted. Ours was the only flight to arrive that morning. No staff. No passengers. We were met in the arrivals hall by our enthusiastic tour guide and herded on to a waiting coach for a 'Tour of Iceland's Magnificent Volcanic Landscapes'. Iceland's volcanic landscapes may well be magnificent but it was hard to tell as they were still cloaked in pitch-black darkness and enveloped in dense fog, punctuated by occasional disconcerting clouds of steam emitting from the roadsides. For all we knew we could have been driving around the access roads of the local industrial estate for an hour.

Iceland was not icy. In fact it wasn't even cold. There wasn't a single snowflake to be had. It had been colder back in Middlesbrough. The 'land of fire and ice' was something of a misnomer; 'land of fog and tarmac' just doesn't have the same ring to it.

The bus came to a halt and we were invited to disembark, to experience 'the raw beauty of the landscape'. I clambered down the coach steps and the icy rain hit me in the face like a jet from a hosepipe, instantly dissolving my mascara and removing three layers of skin, penetrating my waterproofs in seconds.

I found myself on the edge of a cliff, at the end of the world, wild waters crashing on the black basalt beach below me, battered by winds so strong I could barely stand. Father's face had 'WHAT THE HELL KIND OF BIRTHDAY PRESENT IS THIS?' written all over it. And how we laughed. This was insane. Wonderfully insane.

One little-known fact about Iceland that our guide had apparently forgotten to tell us is that nobody actually lives there. It's an illusion, a film set, a beautiful practical joke. The capital, Reykjavík, was completely deserted at one o'clock on a Saturday afternoon. Shops, restaurants, streets – all empty. We wandered through this tiny city, among pretty little wooden houses painted in shades of red, green, blue and ochre and saw no-one. We gazed in awe at the soaring origami-fold walls of the moonstone-coloured Hallgrimskirja, eerily silent and ostensibly abandoned.

Where WAS everybody?

After a brief mooch around the capital, we headed to the famous Blue Lagoon thermal pools for an hour's relaxation before our flight home, and the highlight of the trip. These other-worldly steaming milky turquoise waters are actually manmade, the by-product from an adjacent thermal-powered energy plant. And that's when we found them – the entire bloody population of Iceland, revelling in their glorious couldn't-care-less Scandinavian nakedness, crammed into an open-plan changing room. Mother raised an eyebrow, then muttered something about 'when in Rome'…

Moments later I sank into the hot mineral-rich waters, rain stinging my face. I have never experienced anything else quite like it, before or since. Quite simply it was delicious. You have to go there.

Later that same evening I sat at home with a mug of tea, watching News at Ten, ruminating with some incredulity on the day's events. Did that actually happen? Had I really just been to Iceland and back in time for supper? It may have been the briefest of encounters, but I was hopelessly, crazily and madly in love with the Land of Fog and Tarmac.

***Jane Gulliford Lowes** is a non-fiction author and lawyer from County Durham. After her first travel article was shortlisted in Bradt's 2016 Travel-writing Competition, Jane was inspired to abandon law to write full time. Her debut book,* The Horsekeeper's Daughter, *was published in 2017; her second,* Above Us the Stars: 10 Squadron Bomber Command – The Wireless Operator's Story, *in 2020.*

WORK IN THE 40S

Suzy Pope

Longlisted 2014 'Meeting the Challenge'

The only sign of life was birds of prey circling high above the remains of a cow, its skin in pools around a skeletal frame. I sat on my rucksack somewhere along the border between the Northern Territories and South Australia. Months of picking fruit, endless weeding and clearing gutters left me thirsty for a bigger adventure. The advert in the WWOOF book said De Rose Hill used quad bikes to herd the cattle and I fancied myself a modern-day cowgirl. The wail of crickets rose with the heat. Faint tyre marks in the sand stretched to the shimmering horizon where a silver car winked through a mushroom cloud of dust.

'You must be my volunteer.' A woman with cropped hair, floppy cowboy hat and sunglasses that glinted like beetle shells stepped out to greet me.

I nodded.

'Well, I'm Barb and you'll meet Rex later,' she said, her mouth a tight line. She peered over her sunglasses taking me in; a weedy Scottish girl who looked like she'd never seen a steak before, let alone worked a day on a cattle station.

Vast sheds, old trucks and piles of metal rusting in clumps made up the yard, rising out of the otherwise barren landscape. I scuttled after Barb to the main house. The living room came into focus; seventies décor and pictures of cows everywhere.

'I've ridden a quad bike before,' I said, breaking the silence as Barb rummaged in an old shed full of every shade of rust under the sun.

'Have you now?' she said, dragging out a ladder. 'Well, Rex does the herding, so I guess it'll be clearing gutters and mucking out sheds for you.'

My heart sank.

The burning metal of the gutters made my fingertips raw as I scooped sun-crisped leaves and let them flutter like drunken butterflies to the ground. The heat climbed into the high forties and sweat rolled down my forehead, stinging my eyes. Stringy saliva rolled around in my mouth. Once I'd finished the gutters I was desperate for water.

'What are we going to do with her, Rex? She doesn't look like she can lift a fly, never mind cattle fencing,' I heard Barb's monotone through the open window. A grunt came as a response.

I held my breath as I clattered through the fly door. An old man with a scraggly beard and coat-hanger frame leant against the sideboard. A saggy grin hitched up the corners of his mouth.

'Can I have some water?' I asked.

'Don't take too much,' Barb replied, pointing to a jug in the corner.

It was lukewarm, but I didn't care. I wanted to tip the whole jug over my head. Barb watched with narrowed eyes as I gulped and Rex's grin seemed like it was frozen in time. Neither spoke. Perhaps they hadn't spoken to another soul in years.

'What should I do now?' I asked.

Tractors, ploughs and quad bikes covered in a thin layer of red dust stood silent inside the vast shed. Rex handed me a broom and I expected him to mount a quad bike and power off into the distance. Instead he shuffled to a rusting deckchair in the corner. An old oil drum was his picnic table. A jug of frothing milk that still smelled like cows and a bottle of rum were his lunch.

As I swept, Rex made vowel sounds and pointed at invisible patches of dust. The hot throb of lower back pain pulsed through my body and swirls of dust made my lungs raw, but I was desperate to prove myself. Darkness descended outside, though it was still hours before sunset. Through the open door I saw the landscape tinted an eerie yellow as lumpy grey clouds built up over the sun. A red mist hung in the distance. Rex creaked to life as if someone had wound him back up. He dragged his deckchair outside to watch the sky. Change tingled in the air, or perhaps electricity. A fork of lightning licked silently through the grey clouds.

The wind picked up in seconds as the storm got closer. Flecks of dust grazed my eyes as the world turned red with sweeping sheets of sand. Somewhere out there, Rex sat alone in the storm.

'You alright?' came Barb's voice from behind me.

'Yeah, but Rex…' I started.

'He's not what he was ten years ago,' Barb sighed. Her eyes shone as she stared out into the storm. 'We can't do it by ourselves anymore.'

I didn't know what to say, but an entire desert of sand covered the floor so I cracked my back and started sweeping again.

'Urgh, bloody dust,' Barb said, wiping her eyes. 'Maybe tomorrow we'll get you on a quad bike.'

***Suzy Pope** has taken the plunge into freelance travel and food writing. She's the Food & Drink Editor for* The List *magazine in Edinburgh, has been published in several Bradt anthologies and has written for* The Guardian, Culture Trip *and various hotels and airlines. In 2016 she won a holiday to Madagascar with Hilary Bradt and will always remember the sound advice: 'Don't look directly up at a lemur in a tree with your mouth open.'*

ALPINE ELIXIR

Susan Gathercole

Longlisted 2015 'Serendipity'

Have you ever mistaken your digestif for your aperitif? We did. Two years ago and we are still mortified by the memory.

Our small Cambrian terrace is a lively cosmopolitan place to live, with people of Estonian, French and Romany extraction, not to mention the usual Welsh-speaking, Irish, Scottish and English mix. As with all communities, food and drink is part of the common language helping knit us together.

Gifts of provender often pass between households – fresh mackerel, gluts of beans and berries, new-laid eggs, a bottle of Purple Moose beer from Porthmadog, homemade jam, sloe gin, a jar of membrillo from a Spanish holiday, windfall apples on the doorstep. Small, resonating gestures saying many things – welcome to the terrace; Merry Christmas; get better soon; thank you for feeding the cat or watering the plants; or just – thinking of you.

So when we told Christophe at No. 1 that we were planning a walking trip to the French Alps, he said we must try a special liqueur from his part of the world. He spoke quietly, but with a certain emphasis about this elixir, and when he said its name, Génépi, a romantic longing touched us, stirring the desire to travel south towards the warmth and the artemisia plant that gives the drink its name.

We hadn't ventured abroad for over ten years, so were a bit daunted and shy to begin with. After the packing and passports palaver, soon a thousand miles were under the wheels and we were switch-backing our way up high. Over the landmark mountain pass, the Col de Lauteret, towards the medieval fortress Briançon, and on higher to Ailefroide where the air is washed with a faint blue shimmer from the glaciers.

Towering chamois-studded cliffs surround the village, and we sat feeling very small and dusty but liberated, on a café terrace drinking vervain tisane and gawping up at clouds curdling over the mighty escarpments. At night, fathomless green dreams played out under a net of stars stretching across the dark valley.

Waking to hot sleepy sun and pain au chocolat for breakfast. Lacing up friendly old boots, stuffing cheese and baguettes into the backpack, we set off along dappled birch paths. It was hard to make much progress because the trail is embellished with butterflies, orchids, pasque flowers, alpenrose and strawberries, and each must be exclaimed over and closely examined. Resinous pines perfume the way, and we were soon euphoric from breathing in the warm volatile essences.

Out into a bright boulder-strewn valley. Thundering milky glacial meltwater torrenting past and up we go among the recent violent avalanche debris, shattered rocks and trees strewn across the way. Gasping in fresh cold air blowing down off the snow-fields, we work up and up.

Then a tremendous, fearful rumble from above and a shout from Alex: 'Look up there!' And a moment of terror as a slow-motion snow and ice avalanche tumbles and spews out hundreds of metres above. Scrambling, running, stumbling away towards the other side of the valley, we soon realised we wouldn't be anywhere near the fall, as it drifted down vertically, settling on the pyramid below the cliff face.

Hearts hammering we pressed on again, across a wide rocky plain. Slippery and awkward crossing deep snow patches, then finally towards the valley head where we staggered, shattered, into the

Refuge. There was a warm welcome from our hosts, kindly guiding us to our old-fashioned wooden chalet room, on the way, pointing out marmots through the windows.

When we return to the bar for an evening meal, we are offered an aperitif. We glance at each other, recognising our moment of fortunate happenstance, the benign ghost of Christophe's youth looking down on us encouragingly, and we say gratefully, reverently and confidently yes, we would like a Génépi *s'il vous plaît*.

Time flickers through an icy lens and slows down. There is a silence. Our hosts look enquiringly at each other, then they gently break the humiliating news: 'But Génépi is not an aperitif, it is a digestif.' Oh. *Mon. Dieu.* We blushed, we swallowed, we stammered, we silently prayed for Christophe to guide us in our gauche, blundering Britishness. We accepted the kir proffered instead.

The other guests and our hosts had all said goodnight by the time we finished the spaghetti bolognese and tarte aux myrtilles, washing down our faux pas with a strong glass of red wine. There were a few embers left in the wood stove.

And then the chef unexpectedly emerged from the kitchen, bearing a mysterious bottle, and ceremonially poured us two large shots of Génépi on the house. We raised our trembling glasses to Christophe, and drank the pale green essence of summer in the French Alps, distilling all our romantic longing in a perfect moment.

***Susan Gathercole** was born in Glasgow. She followed her mother's footsteps and attended Liverpool Art College. Later, she turned her hand to writing, as* North Wales Chronicle*'s community news editor. Finally, she built a wooden hut in the back garden, from where she paints pictures featuring colourful ceramics, textiles and people, inspired by memories, dreams and travels.*

WILD GARLIC

Susanna Thornton
Finalist 2015 'Serendipity'

I was in the back seat with my brothers, squashed in next to a battered yellow frisbee, an orange beach ball, an old wooden cricket bat, and a case of plastic boules. There were Ordnance Survey maps jammed in the seat pockets, and a pink bucket full of pebbles and sea-smoothed shards of coloured glass wedged by my feet.

Suddenly the car braked to a halt.

'Look,' said my father. 'What's that?'

It was evening, and we were heading back to the holiday house that we'd rented in St Ishmael's, where the far southwestern edge of Wales slopes down to the azure and glitter of the Celtic Sea. Our car had been nosing along the little road between open fields baked warm from a long day of sun, under the wide sky of Pembrokeshire, with the swish of the roadside bracken and bramble hedges in our ears, and the summery scent of grass blowing in through the open windows. Now we'd plunged into a wood. Dad had been tooting the horn twice each time we came to a corner. Mum was in the front seat. She had leaned back her head and closed her eyes. When the car stopped, we were in a narrow lane overhung with dense trees.

'There,' said Dad.

We all looked. In a stone wall at the roadside was a gate. There was a board next to it. It was dark in the wood, but we could make out a name on the board. *Monk Haven.*

'Let's go see,' said Dad. He turned off the engine.

I was eight and it was August, 1976. Normally we went to North Wales for our summer holiday, but this year we'd driven all the way to the south, the longest journey we'd ever made. We'd played cricket at low tide, we'd dammed streams, we'd raced in and out of the sea, spanking the hard sand and shallow water under our bare feet, splashing, pumping our knees up high not to stumble in the foamy waves. We'd dried our swimming costumes on the warm rocks, eaten salmon-paste sandwiches and wolfed down homemade vanilla buns out of Tupperware boxes. We'd touched red anemones in rock pools and watched them flinch. We'd been to see chapels, headlands, castles. Mum had taught me wildflowers. Ragged Robin. Lady's Smock. Herb Robert.

Now that I look back, I realise that as we drove home that evening, Mum was tired, and probably worried about what we were all going to eat for tea, and whether there was any milk left, and whether any of us had any clean clothes, and how much the petrol was costing.

'It seems private,' said Mum in the sudden quiet.

'Just a quick look,' said Dad.

We got out of the car and walked through the gate. Mum followed my Dad. We found ourselves on a path between deep banks, in a tiny hidden valley. No-one. It was quiet. The cool shade made the hairs stand up on my bare arms and legs. Little dunes of sand piled up in my sandals and shifted under my toes as I walked. A strange smell hung in the air.

'Wild garlic,' said Mum. 'Look.' She bent to show me, and rub the leaves. The valley twisted on. Pale moths flitted in the gloaming. I imagined monks in dark habits, draggled a little and salty with sea water, ropes at the waist, strong bare feet on wet sand. A blackbird spurted out a sudden song from somewhere in the dark branches overhead. Maybe a hermit would be there, alone in a sea-echo cave.

Suddenly the path opened out on to a tiny beach.

'Oh,' cried Mum, softly. We were in a place of wet stones and shingle and shade, a secret cove. The smell of salt and sea and seaweed. We stepped on tangled eels of wet bladderwrack. We slid on green stones and slippery sea lettuce, and our sandals dipped in mud-floor puddles, where little brown crabs tickled out of sight. I walked to the sea and crouched on a rock, and dipped my fingers in the water. Then I rested my head on my knees, and looked sideways at the cliffs, and heard the sounds of the sea change. There were no monks. Just the sad gulls crying, the lilt and flip of small waves breaking. The whole tiny cove in deepening shadow. I walked back. Mum was standing next to my Dad. The light was fading over the calm sea.

Susanna Thornton *was brought up in Stockport. Now based in London, she works in documentary film production, and runs her own YouTube channel about slow travel, mainly cycling and wild camping. Now in her fifties, she has enjoyed travelling (slowly) all her life, doing long journeys such as Hong Kong to London by bike, as well as short, local trips with her trusty two wheels and tent.*

THE SKY BENEATH

Angela Barber

Longlisted 2020 'And That's When It Happened'

Looking out of the window of a jumbo jet it is normal to see the clouds, and land peeping out, several miles beneath the flight path. Above, there is a deepening blue reaching up to touch space, a reminder of just how thin the earth's atmosphere is and how much we depend on its fragile existence. But, on this particular flight, there is a sense of disorientation; the land is not where it should be. Sure enough, there are clouds and land beneath, but cast your eyes above the clouds at surreal peaks of sharply contoured meringue and you'll soon realise that you are entering a world where land and sky seem to swirl and become indistinct from each other. This is the approach into Kathmandu, the capital of Nepal, a country where it is normal to live above the clouds with the sky beneath.

I was en route to the village of Nagarkot, some thirty-two slow and steep kilometres from Kathmandu, at an elevation of 2,200 metres. Our minibus climbed out of the Kathmandu Valley, its engine grumbling up the rough tracks into night-time darkness – a darkness that became so deep that I lost all sense of topography, except for the continuous sensation of climbing. The only artificial light streamed from the van's headlights. It was then that I noticed bright flickering lights appearing out of the side window, and I realised that I was surrounded by starlight, above, and also below me, so it appeared. We had journeyed to touch the sky. And I had journeyed to visit a place that had stirred an inner, inchoate desire within me since childhood. I was visiting the Himalayas, and I was not sure if my mind would ever possess the words to express an emotion so powerful it was almost an out-of-body experience. Ecstasy was not what I thought it would feel like; it was haunting and full of silent tears.

'Set your alarm for 4.30am and be out on the hotel balcony for breakfast. Whatever you do, don't miss the sunrise.'

I stumbled into my bedroom late that evening, by torchlight, in a fog of exhaustion – jet-lagged and muzzy-headed. Did I have a greater need for a full night's sleep or to see a sunrise? Doubts cast aside, I set my alarm and duly dragged myself on to the balcony for breakfast next morning.

The village of Nagarkot is perched on a mountainside with one of the most expansive views eastwards, over eight Himalayan ranges, including Annapurna, Manaslu and Everest. And my hotel balcony clung to vertical mountain rock with uninterrupted vistas over the valley to the mountain ranges beyond. The advertising blurb states that 'when sunrise comes, it just might be the most beautiful place on earth.' I clutch a warming coffee and wait.

Below me is a sea of cloud. I had worried that there would be no view because of the cloud cover, and had not understood how high we were above the cumulus clouds. As the sky grows lighter, tinged with purples, blues and pinks, the ghostly forms of mountains in the sky begin to appear – impossibly high peaks capped with wind-blown snowdrift. The guide begins pointing them out, naming them – an incantation: Annapurna, Manaslu, Langtang, Ganesh Himal, Rolwaling, Jugal, Mahalangur… *Of course*, the local people believe these mountains are the sacred dwelling places of gods. Who wouldn't? And then, in the distance, made insignificant by perspective, the indistinct mound of the world's highest point becomes visible – Everest. Chomolungma. Holy Mother.

And then it happened. A shift in time and light. An altered sense of reality and consciousness. The snow-covered mountains become a changing kaleidoscope of iridescence, until the first rays of sun rise between their lowest peaks. We all stand in silence and then, a Chinese lady screams her wild delight and throws her arms in the air. Others join her in a salutation to the sun. As more reserved Brits, we gaze with lumps in our throats and weep tears of a joy we don't fully understand. The scene seems to shimmer and dance, like a ballet of rainbow light in the pit of my stomach. My legs feel weak and disconnected, unwilling to move and powerless. No words. To this day, there are no words.

I later learned that we were 'so lucky… last week the hotel was shrouded in complete fog and the group saw nothing.' Meanwhile, for the peasant farmers who eke out a subsistence-living on the terraced fields below, this is just another normal start to a normal day. Another hard, back-breaking day lived in one of the world's poorest countries, above the clouds with the sky beneath and a Holy Mother watching over them.

Angela Barber *lives with her husband in Whitby on the edge of the North York Moors. Since retiring as a teacher of English to foreign students, she has completed an MA in Creative Writing with the Open University. She enjoys reading, writing and hiking the beautiful Yorkshire moors and coast around her home.*

PINNACLE POSTCARD

Helen Billiald

Longlisted 2010 'The World at My Feet'

I'd never thought about postcard ethics before.

Fortunately I'm getting a crash course in their morals by a six-foot-square Aussie with handlebar moustache. The moustache is magnificent and it's tricky not to stare.

'Look,' he stubs a finger into my flock of half-written postcards 'you've been there?' I peer across to see a landscape of yellow rocks and blue sky pinned to the bar.

Unfortunately I hadn't.

'Go.' He tries to peel it back with a fingernail but it's stuck to the wood. 'They're a day trip, and then, you can send this,' the card came away with a sound of Velcro and he waved it in my face, 'but if you haven't been there – that's wrong eh.'

Day trips in Western Australia are flexible affairs. Living in a state twenty times the size of England has a way of distorting people's views of time and space; it's a bit like the Tardis. But even here, a round trip of 500 kilometres to stare at a bunch of rocks seemed to teeter on the edge of madness.

I'd heard of them of course. The Pinnacles' severe limestone looks have made them WA icons. Then there's the region's wild flowers, who every August repaint their home in sweeps of blue, yellow or pink, something I longed to see. But this is April and five months of an unblinking Australian sun have put paid to such frivolities.

I went all the same, persuading my partner to flee the manicured beauty of Perth. We drove north and found a land of crisped paddocks, machinery graveyards and wire fences. Occasionally there was the shock of irrigated pastures being doused in huge curls of water, while just beyond died the white skeletons of trees. Altogether it was an unfair season for viewing, making an awkward agricultural backdrop feel gawkier still.

The natural landscape rolled in like a reprieve. We passed strips of red-barked eucalyptus, their skin iridescent from recent rain and clouds of banksias whose torpedo flowers had swollen to a deep butter orange. In between them our horizons were filled with the green of the Kwongan (a word worth speaking out loud… try it), a demure heath-like landscape that would bring any botanist to their knees.

Somehow we didn't expect a visitor centre. Nor the entrance kiosk manned by a chirpy map-wielding woman. She recommends the signed walking loop and the four-kilometre scenic drive while cautioning us not to miss the Desert Discovery Centre. I hear echoes of my friend from the postcard ethics night. 'They're turning Australia into a bloody theme park.'

First glance at the Pinnacles and you have to smile, they have the dreamlike quality of a movie set. We stood and watched the flash of sunlight on glass as a few cars beetled their way round a boulder-edged track. The ocean breeze made their progress curiously silent.

Below us lay hundreds of limestone spires in a rolling sandy desert. There were low stumps, half-buried in vegetation, and slender columns impersonating the remains of wrecked temples. There were lonely three-metre tombstones and semicircular clusters growing together like monstrous sea coral. Inevitably one column stood framed by two beautifully placed circular rocks.

The whole landscape was a lesson in erosion. We felt it as the wind tugged at our clothes and fidgeted its way through the sand. Underfoot, larger grains traced dry streams that crunched as we walked.

By the early afternoon, the light was brutal, sandpapering our eyeballs and mangling any perception of distance. We stole inland for an escape, half searching for wild flowers, half protecting our sanity. In the greenness of the park we found orange banksias but little else, eventually flinging ourselves down for a respite.

Lying there I caught site of a speck of red fluff, like a bobble of wool near my foot. I looked round for a source but only green stared blankly back. Curious I swung round to sprawl belly-down on the sand, my nose a handspan from the vital colour. It glistened in the sunshine, wet red filaments bursting between white sand.

Instead of wildflower meadows I'd found a sundew, a carnivorous plant smaller than my little fingernail, busily hailing insects for lunch. It was the colour of flicked blood.

By the time we stumbled back to the Pinnacles, the postcard scene of yellow and blue was gone. In its place the limestone burned rust-orange and black shadows tracked us across the sand. We stood and watched the strange fierce sun fall into the ocean.

As the water flared a single raven, a black sickle, curved its way over my head and called in a long, loud song before side-slipping into the dark. The silence it left behind settled over us like dust.

__Helen Billiald__ is a freelance writer interested in horticulture and the stories behind people who love plants. With a PhD in ecology, she has worked on gardening titles at Bauer Publishing, UK and has produced science education materials at the University of Western Australia. She writes regularly for gardening magazines and lives in the southwest of England.

JOURNEY TO PACHIRA

Carole Hastings

Longlisted 2011 'Up the Creek'

Bones shaken and minds stirred, the pair of us tumble from the coach along with fifty Costa Ricans or Ticos as the locals are known. After a five-hour journey from the misty highland cloud forests, we're knee-deep in pineapples and bananas on the dock at Caño Blanco, gateway to the Tortuguero Reserve, our jungle home for the week. We're jungle lovers and can't wait to get there. The thrill of spotting elusive wildlife watching us from the shadows is almost too much to bear.

The vibe is distinctly Caribbean – laid-back and crazy at the same time. Chaotically, small boats nip and dart to deposit and collect people, parcels and small livestock. The quay positively heaves. Our bags are whipped away and stacked precariously high on to a tiny boat low in the water. It potters off while we wait for our passenger boat to fight its way to the quay. Police mooch about, long-faced with gravitas, weighing everyone up. It's Good Friday and, as is the law here, they've been taping up fridges containing alcohol in bars and shops to prevent its consumption over Easter. What they can't control is the enjoyment of booze bought in advance. Proudly defiant, two couples with elastic spines gyrate to reggae with brazen beer bottles in hand. We buy some fresh coconut juice from a cackling hag, her toothless head bouncing to the beat. We drink in the action. Bumping and grinding, the men drool over their women. The women prove that with sufficient Lycra, 'one size will fit all'.

Our guide, Julio, points out our boat and we scramble for seats on the oversized motorised canoe, relieved to see a sun canopy as the sweat drips down our sodden backs. The vessel lurches violently, bags and belongings slipping and sliding on the greasy deck. With much humour and even more charm, Julio swaps some of the more voluptuous wenches with their snake-hipped sisters to prevent us from capsizing. A request that may have resulted in much offence in the UK is greeted with raucous laughter and much belly and bottom patting. We chug away from the colourful frenzy at the dock towards the deep and dark shades of green of the jungle. My eyes swivel from bank to bank, while howler monkeys swing and screech high in the treetops and laughing children splash and wave from the shallows.

After a couple of hundred metres on the chocolate-coloured river, the skipper pulls into the matted undergrowth. A massive tree has fallen across the river blocking the way to all but the smallest of boats. There's much shouting and gesticulating as our guide tries to track down a chainsaw. The skipper, bright in his whites, pulls his gold-braided cap over his eyes and snoozes. The laid-back Ticos are not perturbed. Young men wander off in search of illicit beers from the shacks tucked in the trees. Women gossip and giggle, passing round pillow-sized packets of crisps. Lap-hopping babies, all dimples and smiles, are kissed, compared and cuddled by total strangers, while their grandfathers discuss how football should really be played. The only gringos aboard, my husband and I, absorb the scene like voyeurs, wishing we spoke Spanish.

Cheers go up as a chainsaw is found, but it is a while before the smell of cut wood and diesel fill the air. The once-gyrating couples glide smugly by in their small boat. One woman is much the worse for wear – legs, eyes and breasts akimbo. Her disappointed lover flicks water on her face fearing an afternoon of passion is lost to oblivion.

Then a wiry Rasta wading in the river, drags his boat along the shallows, his dreadlocks trailing in the water. His buxom wife lies prostrate like a gorgeous Botero Venus, clutching one bug-eyed Chihuahua to her breast while two more, ears pricked and tails high, stand sentinel on her ample rump.

My thoughts drift and my body steams until another cheer goes up and we head off towards our destination. We've missed an afternoon in the jungle, but who cares? Homer's words ring so true, 'The journey is the thing.'

***Carole Hastings** lives in Winchester and writes purely for pleasure. Carole's play* Living for Today *was in the Chesil Theatre's Festival in 2010. She was thrilled to win first prize in the British Guild of Travel Writer's Competition for New Writers in 2012 for 'Cartagena Night'. In 2018 she came third in the Winchester Writers Festival biography competition with 'Leaving Limbo'.*

RETURN TO SORRENTO

Morna Sullivan

Longlisted 2020 'And That's When It Happened'

I'm convinced I was Italian in a past life. I love the pasta, pizza, coffee, gelato, wine, the country, the language and the people. Every time I travel to Italy I'm persuaded further that there could be some truth to my conviction.

Thoughts of a trip to the Amalfi Coast had been percolating in my mind since watching a television programme about Herculaneum. A last-minute offer proved too tempting. Only after I'd booked the holiday, did I begin to hope the hotel on Sorrento's outskirts would be as good as it looked and sounded online.

'Great!' my partner Gary sighed, while we waited for the tour rep in Napoli airport. While I am a seasoned Italophile, this was his first venture on to Italian soil. I knew that while he was excited to explore the area's Roman artefacts, he had not fallen in love with Italy the way I had, yet.

As the coach pressed through Friday afternoon's autostrada traffic, the tour rep relayed the local history, geography and culture to us. Wispy clouds danced in puffs over Vesuvius's crater, a constant reminder of the destruction the volcano has caused for centuries. Luscious allotments, lemon groves, ancient ruins and pastel-coloured buildings welcomed us as Dean Martin belted out 'Volare', drowning battered Fiat Pandas honking their horns as they outmanoeuvred Lancia Ypsilons. As I breathed in the sights and sounds of Campania, I watched the tour rep talking on her mobile phone. She frowned and walked down the aisle towards us.

'There's a problem with your room,' she said in her lilting Scottish brogue. 'The hotel's been on the phone.'

I looked at her in disbelief.

'We're arranging accommodation for you, close by.'

'OK.' I nodded.

'I'm really sorry. I'll come back to you with more details.' She returned to her seat at the front of the coach, continuing to converse in Italian on the phone.

'At least it's close by,' I said.

Gary rolled his eyes in silence. I reckoned he was now wishing we'd gone to Portugal.

The tour rep returned five minutes later.

'It's sorted. It's just for one night. Then you'll be back in your hotel.' She watched our faces intently before continuing. 'You'll be staying in a convent.'

'A convent?' Gary asked.

'It's stunning.'

'A convent?' he repeated.

'There's sea views. We've booked dinner and wine for you, to make up for the disruption.'

'A convent?'

'It's called Villa Crawford,' she carried on.

'I don't believe this!' I said.

'I'm really sorry,' she continued, 'we're doing our best.'

'I just don't believe what's happening. I know Villa Crawford.'

They both looked at me.

'There's a photograph of Villa Crawford hanging in the hall at home, taken when I stayed in the area years ago. It was dilapidated then. In the evenings I sat at the seafront entranced by the mystery decaying mansion, imagining fabulous parties that took place on the terrace in the 1920s. And now I'm going to stay in it!'

Her eyes widened as I spoke, but not as much as mine did when we arrived at Villa Crawford. An elderly nun smiled, traversing a path shaded by olive trees. We climbed marble steps into a cool vestibule, richly decorated with dark mahogany furniture. The once rundown casa had been transformed. I gazed at high-ceilinged rooms with bulging bookcases and oil paintings encased in gilt frames. The ornate clock on the high mantlepiece ticked slicing the silence. French windows in the lounge opened on to the terrace revealing breath-taking views over the Gulf of Sorrento. I abandoned our luggage, finding a white marble staircase cascading to the top of the villa, flooded with light from a long stained-glass window. Along a corridor I discovered a narrow, stone, spiral staircase, used by servants in bygone days. A moustached man stared from a black and white photograph, welcoming us. Staying in the former home of the American author Francis Marion Crawford was not a part of the Sorrento I had imagined I would be returning to this time.

Entranced, we followed the spiral staircase down to the cellar. The white vaulted dining room took on an ephemeral rosy glow from the setting sun. A dark-suited waiter brought the menu, wine list and his recommendations. I quickly translated the selections and we chose. And then it happened. The sun dived into the water, casting pink and orange fiery rays across a dark blue sea. We sipped chilled Lacryma Christi. Mini cheese pastries whetted our appetites as we awaited freshly caught herb-grilled seabass in the airy restaurant. We didn't need to break the magical silence. Gary smiled at me. And I knew something special was happening. He had begun to fall in love… with Italy.

***Morna Sullivan** has always had a love of stories. She is a member of the Coney Island Writers Group, County Down, Northern Ireland and the SCBWI Belfast group. She has won several writing competitions, has had short stories and poems published but is still chasing that elusive publishing deal. Follow her writing journey at: mornawriter.blogspot.co.uk; mornawriter; and mornawriter.*

THE MAN I MET

Marc Jones

Finalist 2021 'I'd Love to Go Back'

In my mind he's desperate. He's keeping the night close and walking with a lean against the bitter wind. The barest light from the old streetlamp catches him in film-noir frame as he walks past the closed-up store, and I'd love to go back and watch him doing it all over again.

Until then, there's a no man's land around my house. It spreads there, created by a virus pandemic that pins this habitual and heart-steered traveller within the confines of home. It's almost a demilitarised zone, through which only the occasional mask-wearer will pass hurriedly and without confidence to pause and share their day.

When I step away from this window to a changed world, I am surrounded by the remembrances of days gone by. The pictures, the souvenirs, the guidebooks on my shelves all give charge to my spirit and enable me to think greedily of the times I hope will come again soon.

I think of where will be first or at least safest place to go, and where to signal loudest and with deepest joy that this imprisonment is over.

For me it's easy. Closing my eyes and letting myself drift, I can conjure the walk on a hot day, through the blissful emerald soul-collage of Central Park, to the glorious wide steps of the majestic Metropolitan Museum of Art in the concrete heaven of New York City.

I can't wait to bathe in that urban soundscape of life going on once more, and in an unmasked and unafraid bustle. No other city has that 'take it on the chin' arrogance so clearly and boastfully tied to a romantic heritage of lovers and art.

I would choose from the food trucks parked so invitingly at the almost impossibly perfect address of 1000, Fifth Avenue. Could it be any more New York? The Manhattan Hot-Dog, with sweet chillies and harsh celery salt make my mouth water just to even hope for.

Street food and museums. That wonderful brash dichotomy of styles that no other city can match, with its easy contrasts of Luis Vuitton and canvas hi-tops, hip-hop at Carnegie Hall or beautiful stepped brownstones nudging neon Times Square.

After that lunch from the gods, I'll walk up the steps into the beautiful and elegant Met. Feeling the glorious chill within as a respite from the city heat.

This will be a trip with a point. After a year in lockdown, and family life through a laptop screen, I want to share the end of my isolation with the master of its image.

Throughout his abundantly creative life Edward Hopper crafted a connection, for all of us who had not experienced it, to sparsity, loneliness and solitude. His work adorns the walls here in this glorious white marble monument, perched proudly alongside two million other exhibits, which, in our virus-free honeymoon future, will be seen again by over six million people a year.

My aim, when I walk through those boulevard-like galleries, will be to find that man I have missed. He's the figure in Hopper's sketch of 'Night Shadows', a simplistic, yet weepingly brilliant, etching of a solitary man, walking past a closed-up corner bar towards a dark, shapeless nothing away from the reach of the almost pointless streetlight.

When I saw it on my first trip to New York, it was with a sense that I were a voyeur. It was none of my business where he was going or who he was; but I could catch a glimpse of someone else's life in the dark: a street corner with only sinister possibilities.

Now I'll see it as something else. The man is me, walking with confidence and with purpose, away from the shuttered locked-down streets, and into the light, not away from it. I want to nod to the sketch with a grateful understanding that not only our interpretations change, but our ability to characterise our circumstances does too.

During lockdown it had been all too easy as a traveller to feel that same isolation, 'prevented' from returning to the world. But I know that all my travels in the past and those in the future are a privilege. To temporarily suffer the inconvenience of being locked in, healthy, warm and fed, is nothing compared to the many who will not have a return to New York to look forward to.

To know that I will breakfast again in the wondrous Carnegie Diner, will take the spectacular Staten Island ferry past Liberty in late morning, and laze over coffee in Bryant Park once more is to know the world is still out there, and while I can't wait to go back, I know it will be a glorious privilege when I finally do.

Marc Jones *writes at the crossroads where travel meets art and music. After time in Asia, he now lives in South Wales, and is studying for an MA in Travel Writing while curating the TravellersWrites blog which can be found at ⊗ travellerswrites.com. He is currently working on two very different projects: a guide to the Gower Peninsula, and a road journey following the roots of the* Great American Songbook.

3
MEET & GREET

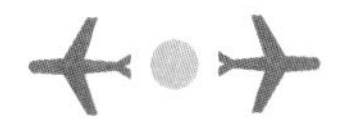

"By the time we pulled into our destination, we had made precious friends and crossed silent barriers."
Robyn Jankel

Memorable travel stories are usually about the people we meet and the interactions we have. Here they are: the funny, sad, beautiful, ugly, positive, negative, neat and messy connections that make us human.

UAE Japan Iran Turkey Kosovo Sweden China South Africa Ethiopia India Arabian Sea Laos Australia Brazil Argentina Mongolia Syria Kyrgyzstan

REFLECTIONS OF DUBAI

Celia Dillow

Winner 2019 'Out of the Blue'

'Why on earth are you going there? You'll hate it.' Friends were not encouraging but I was going to an important meeting.

'Hello,' dark blue eyes locked with mine and I felt reality tilt and reset itself. I was not expecting that.

Dubai takes you by surprise. It is a city where there shouldn't be a city. Some of it is glamorous. It is a gilded building site, a sandpit full of expensive cars and shiny toys. Like a modern paradise garden, everything is reflected in still pools or mirrored tiles. But in the land of mirage, which view is real?

Famous for its skyline, it is all about the building. Think of a shape and build it. Go higher. Twist it like barley sugar and cover it in stardust. Make it glitter. Dream the impossible and make it happen. Put a ski slope in the desert, with real snow and penguins – no problem. Build the earth, or the solar system, in the Gulf and sell it off – done. Bring the nations of the world to this sliver of ancient desert to gawp and gaze and get rich. In that shuddering region, see how safe and tolerant it really is. Defy politics; defy geography; defy belief. Create rain, air-condition the bus stops and put phone chargers in the palm trees – of course. There is plenty of oil but no fresh water, so let the people drink the sea.

Come and party! Move from beach to bar to bed. Ignore the Arabian Desert which licks at your heels, threatening to reclaim the city with every sandstorm. The Dubai Museum tells another story, of nomadic herders, gold traders and pearl fishers who made quiet lives in the *empty quarter*. Simple homes had wind towers with canvas sails to catch the breezes from the creek; the desert was able to support some life. But the medieval settlements have been overwritten in a generation. Out of the blue of the Arabian night, a futurescape has materialised. The wooden dhows on the creek are for tourists now, not fishermen. Celebrities, engineers, financiers and entrepreneurs have come. And they need teachers, hoteliers, diplomats, doctors and nurses. The pace of change is a whirlwind which slices and splices a thousand and one lives into a surprising and uncomfortable storybox.

'I'm just going outside to warm up,' I stepped on to the tiny balcony of a modest apartment to escape the fierce air conditioning. There was the hum of traffic from the highway, the soft stew of the drains and the heat, of course, like something solid. But there was also the sound of half a thousand hammers from the building site below. Hundreds of square feet of pool deck were being landscaped. Sand was shovelled, palms were planted and pools were tiled in every shade of blue. I watched the walls being prepared for the rolls of mosaic. At the beginning of June, it felt dangerous to be outside during the middle of the day; the pavements had melted my shoes. But more than a hundred men, in blue overalls, worked on regardless. Clink, clink, uncomfortable clink.

Rumours are rife. Ten years ago, when the money stopped flowing, Dubai's development stuttered. If the oil runs out, or we stop needing it, how will they keep the desert at bay? If the earth continues to heat up, life in the futurescape will be intolerable and the people will retreat. What will happen to the glittering toys in the sandpit? Will the desert gradually creep back when no one wants to play anymore?

'This is more like it,' it was dawn and we were at an oasis outside the city. Birds gasped in the buzzing heat. Sand grouse hid. There was the flicker of shrike and a dazzling bee-eater. Flamingos gathered on shimmering pools and, not far away, the tawny dune rippled into the heat haze. Natural Dubai pre-dates us and it will outlive us.

I felt splintered by impossible Dubai. My expectations had been overturned. It is a place of hope and no hope, of spinning ideas and spiralling dreams, nightmares and fairy tales. It reflects the best of us and the worst of us.

'Goodbye,' the meeting was over and I staggered home to a new version of normal. At the beginning of the desert summer, in one of Dubai's glittering towers, my first grandchild was born. So, I too have joined the storybox of that sapphire city. Connected now, in a most unexpected way, I left a bit of my heart there. They said I would hate Dubai, but I will go back very soon.

Celia Dillow *is a teacher during term time and a traveller, writer, birdwatcher and hiker during the holidays. She has a PhD in Education from the University of Sheffield. Itchy-footed, she has lived in nineteen houses, in four countries (including Uruguay, Argentina and Italy), and on two continents. Her early travel writing was in actual letters to the people she left behind.*

ODOROKI

Amanda Huggins

Longlisted 2015 'Serendipity'

Sharks and rays swim alongside turtles in an impossibly blue ocean beneath my feet, dinosaurs roam the Jurassic skies above me, and scenes from alien worlds are captured in miniature within glass cabinets and Perspex bubbles. But this isn't a zoo or a museum, it's another of Japan's unending surprises; the eccentric and enchanting Kaiyodo Hobby Train. The carriages are decorated with scenes of sea life and prehistoric creatures, covering the floor, ceiling, seats and curtains in swirls of vivid red and blue. This tiny two-car train is transporting us, and its permanent display case of model toys, through the mountain valleys of Shikoku between Kubokawa and Uwajima.

The train is a travelling advertisement for the Kaiyodo Hobby Museum in Shimanto, home to 10,000 toy figures. It's so delightful, and so unexpected on this rural line, that I find myself grinning spontaneously from ear to ear. Our carefully planned journeys – where we're certain we've left nothing to chance – are constantly catching us unawares. The Japanese would describe this glorious serendipity as '*ureshii odoroki*', which translates as 'happy surprise'.

We've taken cakes onboard for a late breakfast – large doughnuts purchased earlier at the station. We settle into the red velvet seats and take our first bite, anticipating a sweet filling. However, our taste buds turn somersaults when we discover that inside each doughnut is a hard-boiled egg encased in vegetable curry. This wouldn't have been my first breakfast choice, but after the initial shock has worn off we both agree that it's exactly what we didn't realise we wanted. We have accidentally picked up *karepan*, a popular snack that is also the inspiration behind one of Takashi Yanase's anime characters. The renowned artist gave his fictional hero, Karepanman, the ability to burn his foes with the hot curry sauce concealed in his head. Luckily we won't be needing that particular superpower on today's trip, as the dinosaurs and sharks onboard are definitely inert.

Our journey to Uwajima is a two-hour meander through the wooded valleys of Shikoku to the island's west coast. We follow the route of the winding Shimanto River, tilting around tight bends that tip us suddenly into mountain tunnels, and occasionally overtaken by graceful cranes, their black-tipped wings lit by the sun. As we pass through villages we glimpse temples and shrines on the hillsides, and terraced graveyards packed with slender memorial stones. We slow down alongside a man pruning his bonsai trees in the warmth of the autumn sun. He chats to an elderly *henro*, a Buddhist pilgrim who has stopped at his gate, easily recognisable by his conical sedge hat and the traditional white robes worn by those undertaking the eighty-eight temple pilgrimage.

The frequent tiny stations are often nothing grander than a simple wooden shelter and a narrow platform, and at each stop the train collects and deposits glossy-fringed schoolchildren glued to their smartphones, pensioners tucked behind newspapers, and grandmas carrying bags of *daikon* and cabbage. There are intervals of chatter and silence, and one of the younger women practises her English by shyly asking where we are from.

Halfway through our journey, at Ekawasaki, a woman joins the train selling tea and snacks from an ingeniously adapted pushchair.

We exchange bows and names, and buy chilled green tea. Mitsuko points out scenery in incomprehensible Japanese, urging us to dash from one side of the train to the other, giggling at a narrow bridge without railings that spans the river below. We grin with enthusiasm, although it appears tame compared to the vine bridges we have seen in the Iya Valley, their rough-hewn planks separated by gaps as wide as giants' feet.

We nod and smile at mountainside temples, miniature paddy fields glinting like emeralds, trees weighed heavy with *mikan* oranges, and rice straw drying on splay-legged racks, creating the illusion of huge shaggy beasts.

When our self-appointed guide is called back to her tea trolley I turn my attention to the guidebook. I have already learned that Uwajima is famed for its reassuringly bloodless bullfights, where bulls engage in a little macho posturing and then lock horns until one of them surrenders. But now I discover there's a surprise attraction at the Taga Shinto fertility shrine: an eye-wateringly large wooden phallus. Apparently the two-metre-long carving is carried aloft around town on festival days, and I decide it's worth a voyeuristic peek.

As I plan an itinerary, Mitsuko returns to offer us a gift of sweet bean paste balls. I examine them warily, but there is no room to conceal a boiled egg inside them.

'*Oishii odoroki!*' she says. 'Delicious surprise!'

I take my first bite, and Mitsuko nods encouragement, little realising that she has just described Japan in two perfect words.

***Amanda Huggins** is the author of the novella* All Our Squandered Beauty, *plus four collections of short stories and poetry. She has received numerous awards for her travel writing and fiction, and her debut poetry* *collection,* The Collective Nouns for Birds, *won a Saboteur Award in 2020. Amanda grew up on the North Yorkshire coast and now lives near landlocked Leeds.*

EVIL EYE IN ESFAHAN

Sophy Downes

Finalist 2016 'A Brief Encounter'

'Let me tell your fortune,' Nazanin says, unexpectedly. We are sitting in a room in eastern Esfahan, among the paraphernalia of Iranian suburban life: carpets, dainty tables, miniature tea-glasses, and a huge satellite television screen. Nazanin is the *zanbarâdar*, the wife of the brother – Farsi is precise about family relationships – of my indefatigable language teacher, Maryam. I am here to spend the evening in the house where Maryam's parents live with their sons, their sons' wives, and their sons' children – each family on a different floor, like so many chickens in a well-appointed hen coop.

Fortune-telling is ubiquitous in Iran. The night I flew into Tehran we stopped at traffic lights at 4am, and instead of hawking tissues or trying to clean the windscreen, a vagrant sold me, for 10,000 *rial*, a thin, green, printed envelope. Inside was a verse of Hafez – the fourteenth-century mystical poet – like a pearl hiding in an oyster's mouth at the bottom of the green sea. It is called *fal-e Hafez*: you ask a question and then open the envelope. You can have your fortune picked out by parakeets in garden tea houses or find it simply by opening a volume of Hafez in a book shop. All Iranian households, certainly, as well as the Qur'an, own a copy of Hafez which they consult in moments of crisis. But the divination is done through the poetry – it is unusual for people to offer to make predictions themselves.

I met Nazanin half an hour ago. Dressed in a pink jersey tracksuit, she is short and plump, with huge feminine eyes. She is totally unlike me. She has just turned twenty and already has three children – not uncommon in a country where twenty-five per cent of the population are under fifteen. She asks me if I have children myself and if I am married. 'No,' I say, and she commiserates with me: 'Why not? You have such pale skin.' I've only been here five weeks, but I am used to this intrusive catechism and the dubious accompanying compliment. 'Well,' I say duplicitously, 'I have a boyfriend, but we can't afford to marry yet. He had to stay in London to work.' This is not true. I do not have a boyfriend, but I have discovered that this answer meets with approval and that the concept of needing money is understood everywhere. Nazanin nods. She passes me a wriggling toddler and seems to lose interest.

Now, she takes my hand. 'Well,' she says, 'you will have a long life. But your boyfriend – right now he is meeting with a woman, a girl with black eyes and long black hair, taller than you and more beautiful. He is no good for you.' She folds my palm over in her soft hands and gives it back to me.

For a moment I am outraged at this betrayal from my hard-working boyfriend. I feel a frisson of homesickness for distant London before I remember that he does not exist. I give Nazanin a weak smile, but I'm struck by the malice of her prediction. My curiosity flares, and with it an immediate sense of connection.

Suddenly, I want revenge. 'Now I'll tell your fortune,' I say, snatching her hand. I know nothing about palmistry, but that does not deter me. '*In xeili bad ast*,' I improvise. 'This is very bad. So bad I will not tell you what I see.' My Farsi is garbled, but the intent is clear. To my annoyance, Nazanin looks back placidly at me. But my teacher, Maryam, laughs, and then starts to correct my grammar.

Supper is *fesenjan* – chicken sticky with walnuts and pomegranate – herbs, and rice, the crispy *tahdig* crust from the bottom of the pan served automatically to me, as the honoured guest. We sit on the floor. I eye Nazanin warily over the spread cloth, but there is no indication that she is thinking about me at all. Why would she make such a bitchy comment? Jealousy? Pity? Disapproval? I try to see myself through her eyes: blonde, eccentric, old, anomalous, wealthy, free? Does she identify with this mythical black-eyed girl? Maybe she really is psychic, or thinks she is. None of my Hafez fortunes so far have been particularly accurate, but then none of them have had this personal animosity.

The evening ends early, and we go our separate ways. I return to my university hostel and, eventually, back to London, where I encounter no more predatory black-eyed women than usual. But among the many friendly and erudite conversations I had in Iran, this exchange stands out. When I think of Nazanin, it is with an emotion that is curiously affectionate, reminding me that dislike is another form of human intimacy.

***Sophy Downes** is a classical archaeologist, who specialises in Achaemenid Persia. She grew up in Cambridge and now lives in Rome, where she teaches archaeology by day, writes by night, and stalks the Romantic poets whenever possible. She studied at the University of Esfahan and held a fellowship at the British Institute of Persian Studies in Tehran. Her work has appeared in the* Ekphrastic Review *and* Timeless Travels, *and been published by Broken Sleep Books.*

IN THE HOLY CITY OF SANLIURFA

Eithne Nightingale
Finalist 2011 'Up the Creek'

'My name is Hilal, which means good friend. And I hope we will be very good friends.'

Hilal, a Kurd, is our tour guide in the city of Sanliurfa (glorious Urfa) where our group has just arrived on a two-day trip from Cappadocia. He is a tall man with dark hair, a high forehead, thick eyebrows that meet in the middle and a disarming smile.

'Now we will visit the holy fish.'

Hilal is married to a Turkish woman. He is proud of this interracial marriage in the conflict-ridden area of southeast Turkey.

We walk beside the thirteenth-century mosque whose triple domes and stone frieze are reflected in the Lake of Abraham. Women dressed head to toe in the chadar and the burka chatter under the minaret. Many are on pilgrimage from Iran, Iraq and Syria to visit the prophet Abraham's birthplace, sacred to Muslims, Jews and Christians alike.

A veiled woman buys seeds for her sons to feed the portly fish. The young boys squat by the lake, the holy carp swirling at their feet. Then they cry out as a teenager steals one of the fish.

'Bad boy,' mutters Hilal.

A policeman, in a navy uniform and baseball hat, takes chase. The story that those who steal the holy fish will go blind has not deterred the teenager.

We follow Hilal as he strides in his dark suit through the rose gardens beside Ayn-i Zeliha Lake. Zeliha was the daughter of King Nimrod who was furious with Abraham for destroying his pagan gods so had him catapulted from the citadel into a funeral pyre below.

'But Zeliha was in love with Abraham so threw herself into the flames.'

Hilal waves his arms, tracing the fall of the lovers from the hill above. Allah came to the rescue, of course. He turned fire into water and both Abraham and Zeliha landed safely in their separate lakes.

People rush in and out of a row of stone arches ladened with bundles of merchandise. This is the covered bazaar built by Suleyman the Magnificent in the sixteenth century.

'Keep close to me,' says Hilal. 'It's easy to get lost.'

Hilal goes so fast that we have to run to keep up. I am tempted to buy a blue or red headscarf worn by local women. I need protection from the heat and to gain entry to the Dergah Ottoman-style mosque but there is no time. I have to keep up with Hilal. Market sellers peer through strings of dried aubergines and pyramids of turmeric inviting me to smell Sanliurfa pepper but I cannot linger. I have to keep up with Hilal. A handcart piled high with ruby-red cherries blocks my path. A woman buys a punnet to appease her children who spit out the pips at passers-by. I am stuck, lost in a maze of alleyways. Neither Hilal nor the group are in sight. I have no clue of where to meet up or the name of my hotel.

A turbaned shopkeeper points down an alley. I rush past handmade shoes, pots and pans, silks and sheepskins with barely a glance. I turn right into blacksmith alley where old men brandish hot pokers taken out from glowing embers. I turn left into coppersmith alley where young boys hammer intricate inlays on bowls and tea trays, pepper grinders and coffee pots. The sound is deafening. Local women turn to stare at a foreign woman rushing through the bazaar with wild hair and a billowing skirt.

Exhausted, I pass through an archway into a shady courtyard, part of an ancient caravanserai, a custom house, where traders on the Silk Route spent one or two nights free of charge. Turks, Kurds and Arabs sit at small tables playing dominoes, backgammon and chess with tiny brass pieces. There is not a woman in sight. I look round and my eyes glaze over. There are several men with the same high forehead, dark hair and thick eyebrows as Hilal but none respond. They are focussed on their game, their weather-beaten faces creased up in concentration as they consider their next move. Then an Arab with a magnificent moustache, wearing baggy pants and a checked scarf turban stands up and beats his chest.

'I am a champion,' he cries.

I accept tea in a gold-rimmed glass and sit down with a group of men. I am tired of looking for my group. Instead I challenge the Arab champion to a game of backgammon. I hope Hilal is indeed my good friend and will come to my rescue. Or perhaps once again the caravanserai will open its doors for this traveller without silks from the East or a camel in tow.

Eithne Nightingale *is a writer (travel, memoir and non-fiction), photographer, filmmaker and researcher. She writes travel articles for* The Australian *and has won several writing awards including for excerpts from her northern vicarage childhood memoir. She has produced films and is currently writing a non-fiction book about child migration, and recently produced a photography book,* Life Under Lockdown 2020. *See eithnenightingale.com and childmigrantstories.com.*

HANDSHAKES

Elizabeth Gowing

Winner 2014 'Meeting the Challenge'

In front of the mosque, the rubbish heap steamed, and the children crawling over its stench shimmered like visions of angels. It was hard to distinguish anything in the waste other than the forms of the waste-pickers – anything with soft vegetable edges was oozing into a soup of discarded nappies and rotten food, though off to one side I could still see the shape of a dead puppy glistening like a burst fig. Anything with hard edges had already been removed for recycling or reselling. The children were sifting what was left – the trashiest of the trash. Plastic bottle tops were slipped into their bags like coins. A broken toy would be huddled under a T-shirt like treasure.

The nearest girl noticed me through the methane haze and waved. I called 'hello' in Albanian and she stopped. This was a find – a stranger; she clambered down the heap towards me and came to inspect the broken toy of a visitor trying to speak her language. She stuck out a hand, stained and sticky; smeared with something that could have been ketchup from a discarded hamburger, or could have been blood from broken glass. The hand hovered between us for a second while I hesitated, thinking about the dead puppy and tetanus; thinking about basic human connections. I shook it and asked her name.

By the time we were introduced, we were surrounded by other shapes that had come down off the rubbish heap and transformed into children. They giggled and jostled, each child with their hand outstretched to me in greeting, each one with their particular blend of garbage juices across their palms and under their nails. I smiled, swallowed hard, and shook.

That was my first meeting with the Roma and Ashkali community just outside Kosovo's capital, Pristina. I went home and washed my hands in hot water, with soap, and then with disinfectant. I tried to erase the image from my mind, too, of children, who should have been at school, barefoot in the squelch of the rubbish heap. After all, I was only passing through.

When I couldn't wash it away, I went back. I met more children, and was invited into their homes – places made from the hard edges that had been missing from the dump; car doors and flattened oil drums. I met their fathers coming back dirty and tired with the palmful of change they had earned by selling what they'd found in the skips. When I extended a hand in greeting they would shake it with fingers that felt like old leather gloves.

Not everyone worked among the garbage, though – the man I got to know best had previously been a rubbish-picker, but now Ahmet had an office job with an NGO, and his hand that I shook each time we met was as clean as a fresh sheet of paper – but his community still felt like a place dominated by that rubbish heap.

I became squeamish about eating. When I'd shaken hands with all the members of a family in a home with no running water, and their five-year-old came in, triumphant with a golden packet of crisps he'd been given, he gestured 'take one, take one,' but I refused.

It was a family for whom I'd bought some medicine, and the father was angry at my refusal. 'You've helped us so much; please now accept some food from us.'

I wondered whether holding a crisp more lightly in my fingers would stop the transfer of bacteria. I almost dropped it, trying to pass it to my mouth without touching my fingers to any part of my lips. 'Mmmm,' I smiled, wondering what tuberculosis tasted like, and knowing I was shamefully ungrateful.

Over more visits to the community, we finally registered some of those children in school, and I got better at mixing their microbes with my own in the warm press of a handshake. I felt proud of having made these human connections – seeing the hands and not the dirt that stuck to them, making friends with people like Ahmet.

One day Ahmet and I were talking about how some observant Muslims wouldn't shake hands with a woman. Ahmet himself was a devout mosque-goer, 'yet you always shake my hand,' I said. Ahmet looked uncomfortable. 'Actually, when we met and you held out your hand, it was the first time I had ever touched the hand of a woman who was not in my family.'

I thought about all the Ashkali men who had made me feel welcome in their homes, about the sticky, unwashed feeling, the soap and disinfectant, the belief of contamination; and about the triumph of basic human connections, and who it was who'd worked hardest to make them.

***Elizabeth Gowing** (@ elizabethgowing.com) is the author of five travel books, one of them,* Unlikely Positions in Unlikely Places: a yoga journey around Britain, *published by Bradt in 2019. She divides her time between the UK and Kosovo where she is the co-founder of The Ideas Partnership charity which works with the excluded Roma, Ashkali and Egyptian communities who work as rubbish-pickers.*

NAILS OF ICE

Fabian Acker

Finalist 2014 'Meeting the Challenge'

I am above the Arctic Circle where only reindeer, pine trees, and empty Coca-Cola cans mark the landscape. I'm worried about the pine trees. They're like car salesmen, lean, mean and hungry. Their branches point almost vertically downwards, shoulders hunched. They are small, miserable and obviously undernourished. They need a bit of sizzling sun and a steaming hot plate of reindeer dung.

Well they're not going to get either.

The reindeer are keeping all the dung to themselves, probably to provide the raw materials for the carvings the Laps make for the tourists. And as for the sun; well it's summertime and, although the sun is there 24/7, it's invisible. Needle-sharp rain pierces my body, nailing it to the ice.

Apart from the pine trees there are also Laplanders here who live in perfect harmony with nature; the reindeer, the pine trees, the Coca-Cola cans, etc. But like the sun they're invisible. Probably all down at Starbucks. The Laps that is.

No; not true.

Here's one of them; youngish chap accompanied by a blonde-haired lady, both dressed in reindeer skins and beads. (Blonde hair? Lap? A Swedish friend tells me later: 'either the colouring is artificial or she is.')

Oh God, he's going to sing me a yoik. I wipe the rain from off my face, in case they think I'm weeping at the prospect, and let a smile of happy expectation play about my lips. I know that in a few minutes the smile will freeze into position so I stop making the effort and just think of England.

No musical instrument to help him out, but then after a few seconds I realise there isn't one in the world that could. And of course he can keep his hands in his gloves if doesn't have to play anything.

Did I tell you about yoiks? They are songs specific to Laplanders. Sometimes they're tender declarations of love, sometimes they're roistering tales of debauchery, sometimes of noble sacrifice, and sometimes descriptions of train journeys. I think this one might be a love song. Or a train song. Can't detect any debauchery. No; I think it's a song of a reindeer mother separated from her calf who's got her foot trapped in a Coca-Cola can. Never thought reindeers could scream like that.

I try to modify my smile so as to make it look sadly sympathetic to the reindeer, but appreciative of the singer. The whole scene is very sepia. Trees dark against a sullen white sky, a long line of shadowy snow-shoe prints leading to a teepee, which provides the only colour in the landscape. It's made from bright red and yellow heavy-duty plastic; easy to find in a snowstorm. Inside I know there's a gas heater. I want to get in there and make love to it and ask it to bear my children.

I applaud the yoik with just enough enthusiasm to show how enjoyable it was, but without suggesting that I would like to hear another. This is the challenge; can I disguise my absolute hatred of yoiks, rain and ice just enough to avoid offending him and make it to the teepee before I die?

It's not certain. Can't unfreeze the smile. Feet have morphed into ice. Oh God! He's singing another. There is a little chunk of iced water in each ear now muffling the sound but I can definitely hear a reindeer screaming again.

Before he can start a third yoik, I tear one foot off the ground and make a heavy step towards the tepee. I consider engaging his woman friend in conversation to make sure he doesn't begin again. I think of trying 'Do you come here often?', but it's a bit of a daft question with the North Pole just around the corner.

So I turn to the singer. Delicate dark straight hair, fine narrow lines around dark eyes; wind-tanned cheeks from trekking deep inside the Arctic to herd his reindeer.

'I suppose you're outdoors in all weathers, summer and winter.'

'Oh no,' he continues in faultless English as we finally make it to the tepee. 'I don't get outdoors as much as I'd like. My cousin does.' He nods towards a young man sitting down on a tatty armchair. He's listening to an iPod, feet resting on the stove, wearing jeans and a T-shirt decorated with a pair of breasts. 'He's the reindeer herder.'

'And you?'

'Oh, I'm a pilot with SAS.' Pause. 'And so is Monika.'

The girl smiles.

***Fabian Acker** was a chief engineer, a teacher, an editor and a writer. He wrote and edited technical papers for journals in his field and was a talented spinner of award-winning short stories. His story 'Nails of Ice' is a particular favourite of Hilary Bradt's. Fabian died in February 2019.*

THE LETTER WRITER

Joan Waller

Finalist 2007 'A Chance Encounter'

China, for me, is a land full of unfinished stories.

So often, on crowded trains sipping pale tea, at bus stations shivering in cavernous waiting rooms, at airports and in hotel foyers, I chatted to strangers who, knowing we'd never meet again, were ready, even anxious, to tell me their stories. But this time it was different.

I'd walked with Liu Xiao Hua from the university to Post Office Number 2 through the hot, dusty streets of Wuhu. She'd taken on the role of my dutiful daughter – duty in China, she assured me, does not have the cold edge which it has in the West; rather it signifies a blend of love and respect, but I still found it disconcerting, if not irritating, to be treated like an old lady, helped across crowded streets, advised about what to eat and what to wear. Nevertheless, when we reached the post office that afternoon, I was glad to sit back and let her negotiate with the official behind the long counter, fill in the forms, answer the questions. It had been a long semester and I was ready for my summer break.

I sat half-dozing on a hard wooden bench. The air was thick with the bitter smell of cigarettes, noisy with shrill voices and angry arguments. Cloth-wrapped parcels, one end unstitched, were inspected, certified, weighed, sewn up, paid for, thrown into great wicker baskets. People clustered round stone tables littered with grimy glue-pots, fought for stubby little brushes to paste stamps on to their letters, wiped sticky fingers on grubby towels. I'd seen it all before – it had ceased to amuse me.

Then I sensed someone sit beside me, heard the rustle of paper. I opened my eyes. In front of me was a piece of paper, on it, in perfect copperplate, the familiar opening question, 'Do you work in China?' I looked up. A young man was staring at me intently. 'Yes,' I said, 'I teach English at the university.'

He shook his head, smiled, put a finger to his mouth, put his hands over his ears. For a moment I didn't understand, and he pushed paper and pen towards me. I picked up the pen.

'I teach at the university.' I wrote carefully, but my Western scrawl looked ugly beside his fine penmanship. He took his tools back, and I remembered where I had seen him: sitting at a small desk on the steps outside surrounded by a group of rosy, giggling country girls.

Now he started to write at length.

'I am deaf and dumb. I was brought up at a missionary school. The Fathers taught English. I learned about your country and your history. Where do you come from?' Again he pushed pen and paper over to me. 'England near Birmingham.' 'Have you been to London? I would like to see the Thames and your Houses of Parliament. I think they are beautiful. Do you like working in my country?' I nodded, scribbled, 'Yes, I love it here. What do you do?'

'I am a letter writer. I write to the families of peasants from the countryside who cannot write themselves.' I was so engrossed by his story that I didn't notice Xiao Hua had come back. She grabbed me by the arm, dragged me up. 'It is time to go back. We must hurry.'

'But, Xiao Hua…' I was startled; she was such a quiet person.

Then I understood. She'd be in trouble with the university authorities if they knew she'd let me talk to someone with a disability; such people were usually hidden away in prison-like barracks on the outskirts of the city: like crime, disability was something to be ashamed of.

I stood up, smiled at the young man, mouthed 'Thank you.' What else could I do? He folded his paper carefully, tucked it away. Xiao Hua led me out. She never referred to the incident again. For weeks after that I carried with me a postcard, a typical tourist view of Big Ben, the Houses of Parliament, a red London bus. But I never saw him again, even his little desk on the post office steps vanished. It was as if he had never existed. I worried about the young girls who might want to write home: would anyone else help them? But most of all I worried about him: had his disappearance been my fault? Would he ever be able to work again? And once more I was left with a riddle: how did a letter writer who was deaf and dumb communicate with girls who could neither read nor write? It's tormented me ever since; no doubt there's an easy answer, no doubt Xiao Hua could tell me. But I never liked to ask her.

This gentle tale was a finalist in the Bradt Travel-writing Competition in 2007. We have since lost contact with the author.

PONDOLANDAN PIONEERS

Catherine Paver

Finalist 2006 'Taking the Road Less Travelled'

I knew what I wanted from Pondoland. Fun with horses, a cave with no spiders and most of all, to get away from people. When I rode away a week later, my eyes were prickling with tears, but not at all for the reasons I'd expected.

'Pondo people usually see white people from a distance, in their cars,' said Simon, who drove me from the suburbs of Durban to the lush freedom of South Africa's Wild Coast. Pondoland is a small section of it, between Port Edward and Port St Johns. The ride from the Mzamba River to the Mtentu takes you through subtropical forest, red sand dunes and quiet hills dotted with huts.

My guide was a young trainee sangoma called Happiness. All friendly teeth, leopard-print Lycra and brilliant beads, he was a picture of his name. Happiness loved cigars, spoke English well and knew all the important local snakes. 'There is one where we're camping tonight who lives in a tree. He's deadly, but he won't bite you – he's too lazy.'

My dark brown horse was on loan from a Zulu chief. I was travelling with Amadiba Adventures, a community tourist project that uses only local people for every part of the trip. It had just been started by PondoCROP, a resource management and environmental organisation. We would be joined now and again by a few cheerful PondoCROP people, but I was the only tourist, so this was something new for all of us.

The sea was too bright to look at as we cantered along the beach, the wind rushing through the forest beside us. Sometimes we rode into the hills, lulled by the beat of hooves and by the breathing grass. Suddenly, there it loomed on the beach, dark and creaking in the sun: a shipwreck. This was why they called it the Wild Coast.

Shards of rusty metal shifted in the wind, and a sense of disaster haunted the sunlight. 'A few sailors survived,' I was told. They were sunburned and starving, but the Bushmen found them and taught them to live off the ocean. I stroked my horse and tried to imagine them, the first Europeans to see this part of Africa. My ancestors, stumbling through blinding wind towards strangers who might give them life or death.

After a swim we sat on the ship's boiler, which looked like a giant pepperpot jammed in the sand. We munched on the magwenyas we'd brought in our saddlebags. Magwenyas are chewy pillows of sweet pastry and I ate my day's supply in one go. Happiness fetched us fruit the colour of coral from one of the rock pools. It had a milky juice, flesh the texture of raw silk and no English name.

One evening, we were joined around the campfire by Happiness's friend, Wonderful. I'd made friends with tall Tracy from PondoCROP. Tonight she decided to give Happiness and Wonderful a strawberry face pack. She was tired of trying to explain to them what it was for and just made them try it. They rocked with laughter and insisted on keeping the rubbery fronds on their faces, picking at them in fascination.

After supper, we smoked cigars and watched the moon rise, while Happiness and Wonderful talked about snakes. There was one which had a habit of hiding in the roof and biting people's heads, until it was caught by a local woman. She tricked it with a headscarf and a pot of boiling porridge. I didn't find it as funny as they did, but what a bedtime story. 'Who's been exploding snakes in my porridge again?' said Daddy bear.

On the last day, we rode to a Pondo party in honour of an ancestor's birthday. The hut was hot and the ground shook with singing as a man poured a libation to the dead. Happiness was sad when I explained that no, in my country we didn't throw a party to celebrate a dead person's birthday. 'You should! Next time, buy some beer and do it,' he said. There is an African saying: 'People are people through other people.' I thought of it as the voices danced down my spine. Then we rode away. Suddenly I was saying goodbye to people who had helped me, leaving faces I'd remember but wouldn't see again. Human presences stand out so vividly on the road less travelled. They help you to stay alive, and be happy. I had tried to escape, but the laughter of the living and the brave steps of the dead had been with me all along.

Catherine Paver *(catherinepaver.com) writes storytelling songs inspired by her travels in Africa, America and Mexico. Songs to ride away on! Saddle up and listen here: paversongs.com. The people you meet on your travels often make the trip, and they stay in your heart. 'People are people through other people,' goes the Xhosa proverb. Catherine's song about this, 'Way Home', is on her website. She wrote the song in Pondoland! She also wrote sharp, funny columns on teaching for the* TES.

A PRIESTLY APPARITION

Katie Parry

Longlisted 2019 'Out of the Blue'

'We are close now.'

The words have become meaningless. Mohamed, my guide, has been saying them every five minutes for over four hours. I swallow a sarcastic response; his shy smile and widely spaced eyes give him an air of trusting innocence that I cannot bear to disturb.

This moment is symbolic of my time in Ethiopia. The country whispers that beguilement and adventure are just around the corner, but they never arrive in the form that you expect. Up to this point I have spent most of my time either waiting for a bus or vomiting copiously into any nearby receptacle. Today, I have decided, my luck will change. Today I will discover the 'true' Ethiopia – the proud, fiercely uncompromising nation that I have dreamt of during the sweaty tedium of my London commute.

Head down, I continue to slog up the hill. The monotony of the rough, terracotta rocks is punctuated only by pockmarks from the sweat drops that plop regularly from my forehead. The landscape pulsates with the sort of bone-frying, mind-fugging heat that makes you forget that it is possible to feel cold.

A dusty goat, her hip bones jutting from her slender back like miniature copies of the ridge above us, watches us contemplatively. How can she survive here? I haven't seen water, or the faintest hint of green, since we left the car.

Just as I conclude that we must be doomed, Sisyphus-like, to climb this slope for all eternity, we are out of our narrow gully. We stand on the edge of a plateau that rolls up to a dagger-shaped peak. Thankfully, this is not our target. Instead we slide right through a crack in the rock and arrive at a small iron-bound wooden door.

'He is not here.'

I cannot believe it. Have we come all this way for nothing? Too dispirited to speak, I flop into a precious patch of shade and sip tepid, metallic-tasting water.

Suddenly Mohamed lets out a cry, and points back the way we have come.

A dark figure, clad in a long, flowing, white robe, is striding towards us. My rational mind knows that he has just followed us up out of the gully, but I cannot shake the conviction that he is a guardian spirit, manifested in the form most suited to the relentless blue of the sky and the barren landscapes that surround him. He carries a gnarled wooden staff in his left hand, and his greying beard is blown over his right shoulder by a breeze that seems to have sprung up precisely for that purpose.

This apparition is, of course, the priest. As he reaches us, he sinks both of his hands into his robe. With one he pulls out a large, rough-wrought key of exactly the type that should unlock a seventh-century cave church.

With the other he pulls out an iPhone.

'My cousin saw you start and called me.'

His accent is American. I think I detect a faint Boston twang, and feel rather cross. This is not the flowing Amharic of my imagination.

We follow him into the cool of the church, and immediately my churlishness vanishes. The interior is primrose-yellow, and completely covered with biblical paintings. The three wise men ride their camels across the vaulted ceiling in search of a distant star. John the Baptist preaches the good news to a crowd that entirely covers one of the six monolithic pillars. And Jesus himself, his dark eyes almost cartoonishly large, stares down at us from scenes depicting each of his triumphs and disasters.

Reed matting deadens the sound of our footsteps as we wander the aisles. By unspoken agreement we do not speak; the silence is so absolute that breaking it would feel like an act of violence. I focus on absorbing as much as I can and feeling the delicious prickle of sweat drying on my back.

Outside, the priest invites us to share his meal. We squat on our haunches – some of us more elegantly than others – and use the tips of our fingers to scoop up spongy parcels of yesterday's injera bread from the communal pot. The spice hits my tongue with eye-watering force, and I feel my face redden as I try desperately not to cough.

Here, finally, is the Ethiopia that I have been searching for. Few travellers, I am sure, will have made it up to this high, deserted plateau and into the perfectly preserved church behind us.

As we turn for home I notice some oddly straight lines on the ground. They trace the shape of an H. The priest notes my quizzical expression.

'Helicopter pad. George W Bush was here just last week.'

***Katie Parry** is a diplomat who has been lucky enough to live in seven countries on three continents. Increasingly, however, all she really wants is to wake up every morning in a whitewashed fisherman's cottage, ideally in the company of a brindled Staffordshire Bull Terrier. She is writing a book about growing up on a salmon farm in the Highlands.*

MABRAT YELLEM

Joanna Griffin

Finalist 2017 'Lost in Translation'

'Yellem.'

'There isn't.'

I'd learned the word as quickly as the Ethiopian greetings. In a more buoyant mood I'd speak it with acceptance, but it would take on a bitter tone when the small privations of life became too much. Abandoning attempts at conjugation, I'd weave it loosely into conversations with my friends. In our lazy hybrid of English, Amharic and Italian left from the brief occupation, it had come simply to signify 'absent'.

Anything could be *yellem;* rain in the dry season, sun during the rains, milk or meat during the long fasting periods. But mostly it was *mabrat,* the 'light', the electricity. At first, as the dry season had lingered, turning the highlands to dust, there had been no warning. Lightbulbs had suddenly flickered and died and fans had stopped whirring. People would mutter '*mabrat yellem*' before returning to their tasks; little of any importance depended on a power supply. It would return eventually, and I'd frequently wake in the night to a glowing lightbulb above the bed.

For the past few months we'd been on a government 'schedule': a one-day-on, one-day-off approach to the electricity supply, which alternated between the two sides of Gondar town. There was a certain acceptance in simply knowing when we would and wouldn't have power.

But tonight the lights had gone out at dusk and the injustice of it had stung: it wasn't our turn. My housemate, Reiza, and I had stumbled up the track to the kiosk on the corner. Yassin, the owner, was a man of very few words indeed. I liked him for this, and for his uncomplicated acceptance of us, the foreigners in his neighbourhood. Our interactions were simple: I would ask for an egg or a cone of tea wrapped in a page of a discarded schoolbook and he would hand it over with a grunt. No niceties were necessary: *'oncolol'* or *'shay'* was enough. There was a word for 'please' but no-one ever used it.

His kiosk was behind a hole in the stone wall, sheltered by a yellow tarpaulin and usually lit by a single bulb. Above the hatch was an Amharic word in the Fidel script, a series of loops and squiggles I wasn't able to read. I supposed it might have been 'supermarket', a misnomer that graced many a small neighbourhood shop.

Tonight Yassin was standing in the remains of the day, communing with the neighbours and watching the children play. Seeing us approach, he vaulted through the hatch into the blackness of his shop, catching the hanging pens and packets of soap with his heels.

'Uh?'

He turned to greet us with a barely perceptible raise of his eyebrows, the international expression for 'what can I get you?'

Reiza and I turned to look at each other. We must have known the word for 'candle' in this world where darkness was absolute, but it seemed that our memories had failed us.

'*Mabrat yellem*,' I tried.

'*Yellem*,' Yassin agreed, shaking his head. He inhaled as he spoke, in that peculiar manner used by Ethiopians when communicating regret, as though sucking the word back in and swallowing it would somehow mean that it wasn't true. He'd entirely missed my point.

Clutching an imaginary matchbox in one hand, I struck it with the other, and held it against an invisible wick. He stared for a moment before thrusting a small packet of cheap biscuits across the hatch. I shook my head and struck the matchbox again, harder this time with a flourish and a 'zhhhh'. He looked at me without comprehension, proffering some tissues as the light faded further.

Beside me, Reiza sprang into action and leapt into the air, swinging her arms in wide circles. This was 'light', apparently. She jumped and I struck, but we couldn't make ourselves understood.

The spectacle of two gesticulating foreigners began to draw a crowd. Children abandoned their games and gathered round, giggling and shouting objects at Yassin as though it were a guessing game. Teenagers slouched against the wall in casual observation, and women returning from church in their white robes chuckled as they passed.

A youth shambled by and established what it was that we wanted, and a gasp of realisation went through the bystanders as he communicated it to Yassin.

Of course. '*Sham-a.*' Almost the same as the word for 'shoe'.

So, we had light but it might be many hours before the electricity returned, and the thought of lighting a kerosene stove to cook dinner was suddenly a little too much. I turned back to Yassin.

'*Biscoot*,' I pointed to the biscuits with the too-pink filling, still lying on the counter.

'And beer,' I added as an afterthought.

This time he understood. Some things are universal.

Joanna Griffin *has been writing for more than ten years with much of her work being inspired by her time spent living and working in northwestern Ethiopia, her travels around Europe and her passion for open water* *swimming which has taken her to the Arctic Circle, Lithuania, Denmark and around the UK. She has written about wild swimming for Bradt's Devon guidebooks and was delighted to reach the shortlist in the 2017 New Travel Writer of the Year Competition with 'Mabrat Yellem'.*

A PASSAGE FROM INDIA

Emma Channell

Longlisted 2014 'Meeting the Challenge'

This time, I can't hold it anymore: I sit heavily on the pavement and start weeping silently, a local mobile phone vendor staring at me in disbelief.

It has already been three weeks since I arrived. A stone lighter and two hospital visits later, Chennai still refuses to let me in. What on earth could this tall, pale Frenchie hope to find here anyway? Someone asked me that very question once and I couldn't answer. I didn't plan to come, it just happened. I love meat, I wear shorts, and I can't bear the heat. India is not exactly my tailored destination.

Chennai is laughing at me. From the day the plane landed, the city challenges every action I undertake to get to know this beast, this home for billions of people who couldn't be more different from me. I am intrigued and thirsty for discovery, for cultural exchanges, and mind-opening moments. But I remain thirsty for three long weeks. From the rickshaw driver posted in front of my flat with whom I must negotiate every single morning over the same journey as the previous day, to the team working at the coffee shop where I go to spend a couple of hours under the air conditioning in order not to melt, any attempt at meaningful conversation is met with stubborn silence. At the beach, at the museum, in restaurants: silence.

It takes more than that to stop me though, and while I walk along successions of streets where poverty meets with the laughs of kids playing, men chewing tobacco and dogs missing limbs, I formulate a new plan. If I can't get to talk to men, maybe I'll try with women. No doubt that there is much we could share and discuss. A beauty salon is my next step. I am quite happy to mix my mission with pleasure and I start looking for the right place. It takes a while. The sun gets higher in the sky. At least, I assume it does – the intense white humidity makes it tough to be sure. The smell is so strong that I have to cover my nose when passing by the river. I don't like to do this, it makes me feel precious. As quickly as possible I put distance between me and the black water hosting the largest range of plastic bags, scrap metal and cans I have ever seen. A little further, luckily, trees are offering welcome shade and a van sells lime juice. I take one for the road.

I finally find a salon, directed from the street by a long banner portraying a woman with henna make-up and a bindi. The room is on the first floor of a crumbling building with no colour. When I get in, I am pleased to hear that at least one of the employees speaks a little English. I ask for a manicure so I can talk to her. All my questions echo against the wall. There will be no answer. Nor will there be in the next four places I try. It is almost 5pm when I leave the last salon. My hands and feet have never been so pampered. I feel like a joke, a joke with pink nails and orange toes who is going to return home empty of all she was hoping to discover.

And I sit, ready to give up, ready to go back where I come from without getting a feel of what Chennai is. A rickshaw pulls up alongside me. I don't want to be seen crying and sweep my tears with the back of my hand. Stupid pink hand. I look up: it is my rickshaw driver. He smiles at me and throws me the warmest 'lady go house now'. I will even learn his name on the way back: Vishal. But I can call him Vish. And in that instant I realise that Chennai and I are just beginning our journey.

A travel enthusiast, **Emma Channell** *has visited more than forty countries, partly due to her work as an air hostess for Air France in her twenties. After spells in Lithuania and London, Emma now lives in Normandy with her family and still uses any excuse to travel and to write. Fun fact: in 2015, and while pregnant, Emma took part in the first season of* Hunted *on Channel 4!*

CIGARETTES AT SEA

Liz Vernon
Finalist 2016 'A Brief Encounter'

The monsoon was over; *Tokomaru*'s sails hung limp in an unexpected silence. The northeast wind which was supposed to carry us 1,600 miles across the Arabian Sea had died away. Now the sun blazed down on a flat expanse of cobalt blue where not a breath of air ruffled the glassy surface. The sea, once alive with leaping dolphins, flying fish and chittering flocks of terns, had become a desert. Just a few fish lurked under the hull, dorados, motionless in our shadow.

There are two monsoons in the Indian Ocean. Better known is the monsoon which brings strong winds from the southwest and massive rainfall to a parched and overheated India. This one, the one that failed us, brings a steady northeasterly breeze. But it was March, late in the season, and so, hundreds of miles from land we found ourselves becalmed. The diesel cans strapped to the deck were almost empty. The port of Salalah in Oman was 600 miles away; we had diesel enough to motor just fifty.

Then, one morning, I noticed a tiny smudge on the far horizon; we were not alone. Through the binoculars I watched as the apparition slowly took shape and headed straight towards us. Disconcerted, I monitored its steady progress, closer and closer – friend or foe? At last I could see people on the deck – and they were waving! My grip on the binoculars relaxed; it was a fishing boat. They slowed alongside, a battered vessel, not much longer than *Tokomaru*'s 35 feet. The decks were crammed with gear: dark blue barrels, ropes, nets and orange marker buoys. Long wooden booms with pulleys were pointed skywards, as for the moment this crew were taking time off from work, and they were beaming with pleasure.

Whisky! Smoking! they cried.

They were lean young fishermen, five or six of them, relaxed in their faded check lungis, delighted to have company. Before I could think what we had to barter they started to shower us with presents – a coconut, followed by biscuits and packets of noodles; they were clearly well provisioned. I found some dates, mints and a torch to offer in return. It was a friendly exchange; we were much cheered and entertained on both sides by this chance encounter in the empty, windless Arabian Sea.

Loath to go, they tried again: Whisky! Smoking! This time, to encourage the desired response, we were offered a massive yellowfin tuna. Alas I had to refuse, having no freezer. A water melon then? Gratefully accepted. Some chocolate in return? Yes, yes! Thank you!

But 'smoking' remained the goal. Nick, as it happened, had been emptying the last of the diesel from the cans on deck into the fuel tank. He waved the empty 25-litre can – diesel?

Diesel? Of course! In exchange for smoking?

Okay! In exchange for smoking.

Smiles all round – we had a deal. Nick assured me that they carry tons of fuel, they could spare a few litres.

Now it was all action. A rope was thrown across to our boat to attach the fuel can. Two cans! they gestured – two! Having reeled in the cans they cut their engine and started pumping. The sea state was perfect for these antics, two boats, side by side in a flat calm. Time for some introductions. They were from Sri Lanka, away for a week, offshore fishing for tuna and grouper. Arosh and Deepal organised the fuel, while Givantha examined our fishing arrangements. He was not impressed. Before I knew it we had a fine new lure. Meanwhile I prepared our side of the bargain – two packs of cigarettes and some dollars. After all, to have diesel delivered to one's boat in mid ocean is worth more than 'smoking'. The cans come back full, and over went our bag in return. These generous, good-natured young men didn't even look in it, so much fun were they having in rescuing the becalmed sailors.

Time to go our separate ways, they to the east, homeward bound, we to the west, much encouraged, with another thirty hours of motoring in reserve. That night, just as the fishermen had lifted my spirits, so a soft breeze lifted our sails and we began once more to move, yard by yard, across the Arabian Sea.

After forty years teaching in London schools and colleges, ***Liz Vernon*** *retired in 1999. Time off during her working years was spent coastal sailing around Europe with her partner, Nick. On retirement, they bought a 1970s ketch,* Tokomaru, *and headed west to sail around the world. The voyage extended to ten years, ending back on their mooring on the River Orwell in 2009.*

52-CARD PICK-UP IN LAOS

Joanna Mason
Finalist 2015 'Serendipity'

It was on the banks of the Mekong, as I was taking a walk at sundown, that I found the last card. The King of Hearts lay there waiting to be discovered among the dust and an old empty crisp packet, being pecked lightly by the weary-looking chickens that had gathered for this momentous occasion. I now had a complete pack, 52 cards plus both jokers. It had been quite a journey across Laos to find them all.

It was over a few beers and during the most amazing lightning storm I had ever witnessed that I first heard of this bizarre collector's game. I had just crossed the border from Thailand a day earlier and was waiting for the storm to pass before moving on. The American explained, through sips from a Beer Lao bottle, that all over the country you could find playing cards strewn on the floor, in bushes and by the side of the road. Forked lightning lit up the sky, narrowly missing a rickety TV aerial on the tin roof of the next building. 'Once you start looking you'll see them everywhere and you'll be hooked. I'm looking for the Four of Clubs. If you find it, I'll trade you.' In answer to why the country seemed to be sprouting playing cards, he just shrugged and took another slug from his beer. After a long, relaxed pause he did explain that many travellers were taking on the challenge to seek and retrieve a full deck.

And he was right. After I saw the first card outside my guesthouse the next day, half-buried under a pile of sugar cane, I just couldn't stop finding them. I was totally hooked and determined to complete my full deck before I left the country in a month's time.

I sat there on the tan-coloured banks of Si Phan Don watching the last of the evening light dance on the surface of the Mekong River, thumbing the dirty, dog-eared King of Hearts. Some local children were getting ready for their evening bath-time ritual; taking it in turns to sud up their hair and bodies with soap before somersaulting headfirst into the river.

I let my mind wander back to my favourite card and the serendipitous events that had brought us together. I had boarded a local bus in Luang Prabang headed for the pancake-flipping, backpacker capital of Laos: Vang Vieng. The bus had been pre-loaded before the passengers' arrival with huge sacks of white rice. They were wedged under every seat, on the rusty old luggage racks that now bowed worryingly above my head and into each footwell. I sat with my feet on a sack and my knees around my ears and hunkered down for a long, uncomfortable journey.

About halfway the bus stopped. I peered out the window, the limestone mountains stretching out before me, the remoteness of where I was just hitting me. I couldn't see anyone waiting to get on the bus and no one got up to get off. Then, with the stealth of a panther, a man emerged from the wilderness. Dressed head to toe in camouflage and carrying a wooden-handled AK-47 assault rifle. Suddenly realising I was the only foreigner on the bus, I began to panic. The guidebook's warnings of bandits and hijackings rushing back to me. The man's black army boot fell heavy on the step as he climbed aboard and as he looked up his eyes fell directly upon me, my mop of blonde hair standing out like a sore thumb. My mouth went dry and mind began to think the worst. As his eyes met mine, he smiled a wide, toothless grin at me and quickly sat down in the seat directly in front of mine, resting his trusty gun in the empty seat. He peered back at me through the gap between the seats and looked down to the collection of playing cards that were resting in my lap. Quickly his face disappeared from the hole and in its place a sun-toughened hand appeared and thrust a small papery ball into my own. I stared down, not sure what had just happened or what to think. I began to unravel the crumpled package to find a rather dirty, creased Queen of Diamonds staring back at me. At that moment, the worry of my impending death by firing squad disappeared and I lifted my head to thank the man. But I was too late, the bus was already pulling to a stop and the man slipped off and back into the wilderness as gracefully as he had come. Just as he stepped into the leafy jungle, he glanced back at me, flashing me one last toothless grin.

***Joanna Mason** is an aromatherapist who lives by the sea with her family in Hastings. Besides being a mother, she believes she was born to create. Mostly jewellery, but she dabbles in a million-and-one creative practices, including writing. You'll find her now… dancing around her front room with her two girls or with one toe in the sea.*

SAVED BY A FROG

Mike Crome

Longlisted 2015 'Serendipity'

An axle snapping sounds a bit like a bone breaking, especially if it's heard as you drive through the desert on the northwest coast of Australia. An agony of silence follows as our old campervan, suddenly powerless, limps to a dead stop.

Sheila and I have been dreading the possibility of this happening for weeks, as we attempt to take our antiquated Winnebago on one last lap of the Australian continent, a journey of over 16,000 kilometres and one of the world's great road trips. Now, almost halfway around, disaster strikes in the worst possible location. In either direction, the black ribbon of highway disappears to a vanishing point in the featureless landscape of acacia scrub, spinifex and red earth.

After a short sad wait, a passing truck tows us to Pardoo Roadhouse, a dozen kilometres back along the highway, where we confirm the worst – the axle can't be fixed here. The nearest town is 120 kilometres away and a phone call to the Port Hedland Towing Company yields a promise to send help – in three days' time. We spend a long weekend of boredom, frustration and self-recrimination at Pardoo, until our vehicle is finally winched on to the back of a flatbed truck and we make the costly journey to the port.

Our first view of this outback seaport on the edge of the Indian Ocean is not encouraging. The town is defined by a razor-sharp horizon that dissolves into a watery mirage where the brutal hand of industry has transformed the stark wilderness into a post-apocalyptic nightmare, like a scene from *Mad Max*. Port Hedland is surrounded by bulldozed mountains of sea-salt and abandoned heaps of scrap metal; it's fed by freight trains five kilometres long, screeching and growling on tracks beside the highway, their trucks stacked high with dark red iron ore. A tattered windsock blowing in the desert breeze indicates the airport and across a desolate plain of salt, rusty sculptures – Cubist crushing plants, conveyor belts and the funnels and superstructure of freighters – are silhouetted against the merciless blue sky.

At the mechanic's yard our woes are compounded when we learn that it will take a week or more to repair the Winnebago, if they can find the parts. Our first night in Port Hedland is spent in a motel on the wind-blown outskirts of town, our room a converted steel container called a donga, with sealed windows, two springy beds, and icy cold air conditioning. At dawn we check out and take a taxi into the town centre in search of a better life.

Nearly every building is stained in a hellish palette of red primer by the pervasive iron ore dust. Even the footpaths are red, as is the grass on the lawns, the leaves on the trees, and the scruffy pigeons hanging about the main street café. Abandoning all hope, we enter the tourist information centre and learn about a local hostel called 'Frog's Backpackers'. The lady in the office gives the owner a call and a few minutes later a battered ute pulls up and out steps a lanky, red-headed man dressed in a T-shirt, tight shorts and big Aussie work boots.

'Gooday,' he says, 'My name's Frog, hop in an' I'll drive ya to me hostel.'

And so, from the depths of despair, we are taken to Frog's – a cool, tranquil retreat overlooking the turquoise ocean, where, under the dappled shade of palm trees we watch the huge ore ships setting sail for China. Frog's is a place for weary travellers to rest; where friendly folk are drinking cold beer; where I can walk across the road and cast my fishing line into the Indian Ocean and pull dinner from the sea. It's a place where Sheila can sit and read her books and I can strum my guitar and write my story; where we play chess with steelworkers and make lasting friendships. It's all down to one man, Frog, who has created this outback haven, not for profit, but for the pleasure and camaraderie each new traveller brings to such an unlikely place.

In the end we spend three weeks in Port Hedland – the replacement axle is hard to find. Frog's hostel transforms our experience, and we slowly grow fond of the dusty red town, a place we would never have stopped in by choice. I suppose the lesson for travellers is that you will encounter hardships along the way, but you have to make the best of it and with luck, and a bit of serendipity, you might just end up as a guest of someone like Frog. I only wish we could take him with us when we leave Port Hedland, we might just need him again further down the road.

***Mike Crome** is a Sydney-born Australian who lives in Norwich, UK with a partner called Sheila. Together they run their home as a travellers' guesthouse and he paints pictures, writes stuff and plays guitar in a band to make a living. This lifestyle has enabled them to travel extensively over the twenty years prior to Covid. He hopes that we can all travel again sometime… meanwhile, he thrives off the memories.*

THE STANDING TRAIN

Robyn Jankel

Highly Commended 2014 'Meeting the Challenge'

Haerbin's illuminated ice castles were melting.

We were English teachers in China, holidaying in the north. As winter thawed, our Cantonese classrooms beckoned, but Haerbin remained unvisited. Trains were standing room only and the journey would last 24 hours. But the ice castles were melting and so was our time. Such discomfort was fair exchange, we reasoned, and snapped up the few remaining tickets.

At 2am on the platform, surrounded by leathery peasants bearing folding stools and determined grimaces, the first doubts appeared. When the overstuffed train rolled into the station and guards forced a chosen few through its groaning doors, our apprehension increased. Once on board, I surveyed the seething mass of humanity and realised our catastrophic error. Heilongjiang's snowy palaces shrank to a fabled, distant dream. What we had willingly entered was a nightmare.

The train was packed. And not rush hour on the Tube packed, but impossibly, astonishingly, terrifyingly packed; the wooden benches stuffed with a dozen passengers, children crouching under tables, adults climbing on top of one another and an unmoving, solid throng of travellers compressed into the aisle. Yet with no space to breathe and surrounded by groping limbs, the passengers' only consternation was that four westerners now shared the claustrophobic lowest class. Unable to squeeze through into a carriage, I remained in the enforced limbo of the joiner; wedged between the wall and a toothless woman who thrust a bucket into my arms.

We anticipated the first station almost as much as our distant disembarkation, but it was to prove an almighty disappointment. A few passengers escaped but hundreds more tried to take their places. With no entry through the bulging doors, they climbed on to the outside of the carriages. As we puffed slowly out of the station, stooped men and wiry old women lowered themselves through the windows on to the boiling sea of people below.

The night air dropped to -10°C and ice formed on the inside of the window. As my fleece stuck to the frosty wall, I thought longingly of the down-filled coat under my feet. Lashed to my rucksack and relegated to the floor on our arrival, it now served as little more than a lumpy carpet. My head pounded from the frozen air and increasing dehydration; our carefully packed water bottles and essential Chinese train journey sustenance (biscuits, mini jellies, cup-a-noodles) nestled in our rucksacks alongside those longed-for additional layers, mere inches away yet painfully inaccessible. This inability to eat or drink was arguably no bad thing since the tiny toilet had been taken over by a family of three leaving no room to close the door, let alone use the facilities.

A coughing fit offered a brief respite from the crush. My horrified neighbours thought only of the SARS epidemic and managed, impossibly, to back away while I groped blindly for my inhaler. A baby peered inquisitively from the luggage rack.

Some eight hours into the journey, we knew that the train would pass through Beijing. I wondered: should we leave? We could just get off this journey into the depths of a frozen hell and cut our losses. How breath-taking could those ice palaces really be? How much did we care? There were no alternative tickets or methods of transportation. We had chosen the busiest weekend of the year to make this journey, immediately before Spring Festival, alongside hundreds of millions of people simultaneously crossing the country to get home to their loved ones. Wherever we disembarked, that's where we'd welcome the Chinese new year. But we could sit. We could drink. We could breathe!

We stayed on the train.

The hours were painful but as the sun rose, they seemed shorter. The bucket lady handed around kumquats, closing our protesting hands around her precious wares. A man with a puppy in his pocket shared his sunflower seeds. For a brief moment the area was filled not with the laboured breathing of constricted, contorted individuals, but mutual appreciation of shared discomfort. When thirst and sleeplessness caused blackness to cloud my vision, I was thrust ceremoniously and inexplicably into the carriage. Crowds miraculously parted, the edge of a bench made free, a dirty bottle of water appeared from nowhere. Somebody started singing. In the afternoon light, we could see views from the windows. By the time we pulled into our destination, we had made precious friends and crossed silent barriers.

In Haerbin, the ice castles were stunning.

We revelled in their chilly splendour, then sat down and drank tea.

Robyn Jankel *spent a year in China when she was 18; now in her mid-thirties, she still has the travel bug. She is currently working on a book about cycling long distances in high heels and inappropriate clothing. Robyn lives in York where she runs a small B&B.*

THE STREET CHILDREN OF SALVADOR DA BAHIA

Sharon Watson

Longlisted 2010 'The World at My Feet'

I notice them long before they notice me. From my elevated position in a first-floor *por quilo* restaurant in Salvador da Bahia, I look down from my window on to the street below. Two boys sit on the pavement, one perhaps 10 or 12 years old, the other perhaps 7 or 8. The younger child is lying with his head in the lap of the older one, who is combing the younger child's hair with his fingers and gently caressing him. Both are barefoot and shirtless, the younger boy's tight black curls dirty and matted.

The younger boy begins playing with the older one, and they wrestle a little, laughing. The older boy looks up casually and notices me sitting there, watching. Instinctively I avert my gaze. I look back, ashamed at my action. I do not want to avoid confronting Brazil's poverty, especially not in this city.

The eldest boy throws a blue and white sheet around the two of them to protect them from the wind. They laugh and play underneath it, oblivious again to my gaze.

I continue eating. When I next look down at the street, the younger boy has a container with the remains of someone's milkshake in his hands, and is sprinkling it over the head of the elder boy, who laughs good-naturedly and looks up as he brushes the milk out of his hair. He catches my eye again, and this time I do not avert my gaze, but instead smile and laugh. The elder boy grins, nudges the younger one and points upwards to tell him that the lady is watching. He makes a sad face at me indicating they are hungry, and I nod and give him the thumbs up to tell him I understand and will do something. He grins and gives me the thumbs up in return, and I indicate with my finger to tell him to wait while I finish my meal. He smiles and nods.

A squall whips up and the pair race across to the shelter of a shop awning. When it passes they return and I finish my meal and indicate that I will be down shortly. Before leaving the restaurant I take R$10 out of the zippered compartment of my bag and pop it in my pocket, sealing up my bag again. My language skills are not yet at a point where I feel confident about purchasing food from the restaurant for these kids, but I know that with R$10 they will be able to buy themselves a good meal.

I walk outside and hand the eldest child the money, smile and begin to walk away. He takes it, walks away a few paces and then turns back and approaches me again, his eyes shining. 'Obrigado amiga' he says, and embraces me and kisses each of my cheeks. I am touched by this spontaneous expression of affection and smile shyly at him. As I turn to leave, the younger boy comes running down the pavement and points to the sores on his arms, clearly hoping I will give some more money. The eldest child chides him, his years bringing him wisdom, and urges the younger one to come and get some food. They leave together, the elder child ignoring the complaints of the younger.

I cannot change the world, and I cannot fix Brazil's poverty. I cannot even give money to every homeless person who asks me, because there are so many. All I can do is tiny things that make a difference to a single day in a person's life; a few reals to an old lady outside a church; a few reals to a man with no legs, flip-flopping his way around the city with his hands in thongs; a decent lunch for these two children of the street. It is not much, and it does not assuage my conscience.

Postscript: I saw the eldest boy again the next night, and I allowed him to take me by the hand and lead me through the streets to a crowded grocery store. After pushing his way to the counter, he asked the storekeeper for three large tins of powdered milk and indicated I would pay for them. The storekeeper looked at me slightly sceptically, but I nodded my assent. As I exchanged money for the tins, which the boy piled up proudly in his arms, an audible ripple of astonishment flowed through the crowd. Tourists, I gathered, didn't usually do such things. I handed the boy the change the storekeeper had given me, kissed him gently on the forehead and walked quietly away, humbled by the experience. It still was not enough to assuage my conscience, but it was something.

Sharon Watson *is an educational designer who was born in Melbourne, Australia and currently resides in Canberra. She is addicted to travel and music and sometimes combines these passions by travelling to far-flung places to meet friends made through the U2 fan community. Such a friendship led her to Brazil, a country she fell in love with and hopes to visit again in a post-Covid world.*

SHOELACES

Liz Gooster

Longlisted 2010 'The World at My Feet'

If it hadn't been for Senora Zapato, as I'll call her, not knowing her real name, what happened to me a few days after I met her would have seemed even more bizarre. It was late afternoon with a sullen sky and I was drifting down Calle Peru, one of the shadowy, narrow streets in San Telmo. The heavy air was filled, as it is everywhere in the city's oldest barrio, with the constant, coarse symphony of blaring bus horns. I was on my way home from Spanish school, hazily mulling over the impenetrable and daily growing jungle of subjunctive tenses in my mind, when an elderly but sprightly woman, dressed in a smartly buttoned coat with a fox's head peeping demurely over one shoulder, stopped me. It was Senora Zapato. I assumed she wanted to ask me for directions: it happens to me wherever I am, from London to sub-Saharan Africa. I feel pity for the dislocated strangers around the world who seek my help, because I have a limited sense of direction and am often lost myself. Hector, the doorman in my apartment block in Buenos Aires, joked that he was going to put an arrow on the wall pointing towards the front door, because I kept turning the wrong way when I came out of the lift.

Senora Zapato didn't need to know where she was going. Instead, she said something to me in a civil yet urgent tone, undoubtedly polite but to me, still unfamiliar with the coasting cadences of the Argentinian dialect, inexplicable. When I continued to look blank – I was still listening out for a recognisable street or familiar place name on which to hang the hook of my groping Spanish – she began to gesticulate firmly at my shoelaces. They weren't undone. They just somehow didn't meet her standards of safe neatness and had too much waywardly trailing lace. I thanked her for her advice and went to move on, but she wanted to witness me tying my laces tighter, so like an obedient child I propped my feet one by one on a convenient doorstep and retied my trainers under her beady inspection. When they were finally done to her satisfaction, she give me a big thumbs up, an endearing grin – mirrored cheekily by the fox – and swept on.

A curious but unique occurrence, you might think. Senora Zapato was a touch eccentric, you might conclude. Yet a couple of days later, having dropped off my washing at one of the many super-fast Chinese *lavanderias*, where your clothes are in, whirled, tumbled and out in a matter of hours, I overtook a young man on the pavement outside one of the equally numerous, but ploddingly connected, internet cafés. Eager to hook myself up to the IV line of Gmail, I brushed past him impatiently and as I did so he inadvertently trod on the back of my trainer, which slipped off. I shoved it back on and shrugged off his effusive apologies. He carried on talking but my mind was already composing an email to my estate agent in London and I was barely listening. The guy was being very persistent about something and eventually I realised that part of his phone charm, a shiny blue ladybird, had somehow fallen off and become lodged in my shoe. He bent down, pumped his upturned palm in the air until I raised my leg slightly, then wriggled off my trainer and reclaimed the ladybird. Odd, but well, maybe it was his lucky phone charm.

Expecting him to straighten up I turned towards the welcoming doorway of the internet café, backlit by the blue-grey flares of a suite of computer screens. Like a salesman in a child's shoe shop, ladybird man remained crouched on the ground, while he doggedly relaced my left shoe. Then, despite my bewildered protestations, he tapped my right foot, eased off the shoe, shook it interrogatively, tenderly replaced it and retied it securely with a firm double knot. I felt like a bemused modern-day Cinderella, in scruffy Skechers rather than glittering glass slippers. Both shoes now tightly and symmetrically laced, he stood up, said goodbye, and walked off. I stood, mute, my skimpy Spanish having fled in incredulity. I'd heard the warning tales of wily *Porteños* flinging ink on tourists as a distraction while they rob them and when I reclaimed my senses from the soles of my feet, I began to wonder if this was the latest scam. But my purse was in my bag, my watch was on my wrist, my laundry ticket was in my pocket and what's more, I had the best-tied shoelaces in Buenos Aires. I shrugged my shoulders and nudged open the door of the internet café with my foot.

Travel writing blends two of ***Liz Gooster's*** *big passions. The inspiration for 'Shoelaces' came from her time in Buenos Aires during a six-month sabbatical in South America. Professionally, she has her own positive psychology coaching practice. She lives with her fiancé and 3-year-old daughter in Cambridge, where they're currently renovating and building their new family home.*

FRISBEE DIPLOMACY

Gordon Thompson

Longlisted 2019 'Out of the Blue'

Our minibus was a tiny boat in an ocean of grass. Half an hour out of Ulaanbataar, the purple Hyundai began to struggle across the Mongolian steppes, and several times the driver had to pull off, clamber around back, tinker in the engine bay, and climb back in.

'All Mongolian drivers also mechanics,' Uuganta, our smiling, bespectacled local guide, assured us. 'Country is so big, not possible call for tow truck. Driver must know how to fix everything. Our driver is very good mechanic: he fix bus, no problem.'

Anxious looks passed among the seventeen of us. After an overnight train from Irkutsk, Russia, we were desperate for proper beds, food, and drinks. We shared unspoken relief when Uuganta told us, 'I just call office; they send new bus – not possible complete trip like this – take several hours.'

When it became clear that we'd be immobile for a while, Dave got out his frisbee. A British guy just out of college, he was on a mission to photograph himself throwing his frisbee past major monuments. He'd already logged Red Square, and was looking forward to the Great Wall of China. Eric and Conall and I followed him off the road and began throwing the disc around.

Suddenly two horsemen trotted across the road right up to our circle. Garbed in the traditional handwoven tunic and sash, they looked larger than life on their shaggy horses no bigger than ponies.

They dismounted and squatted: an older man, maybe sixty, with a leathery face, and a younger man, his son presumably, with equally tanned but much smoother skin. They lit cigarettes and watched in bewildered fascination. We might have been commanders of a flying saucer: it was clear neither of them had seen a frisbee before.

They drew closer, and we worried we might hit them with the frisbee: at one point Conall grazed one of the horses on the muzzle, but the old man yanked it out of its spook, nodding as if to say, *Don't worry about it.*

'Are you from town?' the old man asked with Uuganta translating.

'He means Ulaanbaatar,' Uuganta said.

'We're from England,' Dave said.

The old man looked baffled even after Uuganta translated.

'The United States,' I said. 'Across the ocean.'

'On the other side of the earth,' the old man nodded. 'I've heard of this place.'

We offered to toss the frisbee to them a couple times before the young guy got up and haltingly joined in. The old man shook his head in bashful awe, the way a child shrinks from touching a magician's hat. It was thrilling to realise there were still people in the world who could be awed by a frisbee, but disappointing to think we were reducing their galaxy of wonders.

'Did you make this?' the young guy asked, turning the frisbee over reverently.

'No,' Dave said. 'But I suppose you could.'

'Yes,' he said. 'With a spring sapling…'

'Beautiful horses,' I said, offering my palm to the young man's bay.

'You want to ride him?' he asked.

I looked around for support.

'Go on,' Uuganta urged. 'Cannot be true Mongolian man without ride a horse.'

Before I had my feet in the stirrups the animal took off, and it was all I could do to keep him below a gallop with rope for reins, a tiny saddle, and my shoes almost scraping the ground. The boundless earth unfurled all around me like an enormous flag. It was a thrilling, terrifying half-mile before I managed to wheel the horse around and trot back to the group.

For the first time, the herdsmen were grinning.

'He says you look like a tree on top of the horse,' Uuganta smiled.

'Maybe I should stick to frisbee,' I panted, dismounting. 'Thank you.' I handed over the reins.

'Come visit us. Just over there,' the young guy gestured. 'Horses for everyone.'

'Wow,' Conall said. 'Could we?'

'Very long walk,' Uuganta said. 'No horses where we stay.'

Suddenly the driver announced that the bus was fixed. The horsemen re-mounted, and Dave asked if he could take their picture.

'Good time to go,' the old man said. 'Bad winds coming.'

The sky was an almost cloudless sweep of blue.

By the time the bus got underway the herdsmen had cantered off, vanishing into the rolling plains dotted with sheep and goats and horses.

Shortly after we got to our yurt camp a beautiful sunset turned into a hailstorm. Nestled in by a dung fire, I looked at the frisbee on Dave's bed, and as I remembered the horse bounding beneath me, the uninterrupted horizon all around, and the enigmatic herdsmen's sudden laughter, I thought that magic has not gone from the world after all.

Having earned an MFA from the University of Florida, **Gordon Thompson** *teaches at Cranbrook School in Metro Detroit. He has also built trails with the Forest Service, served as a Kiva Fellow, and worked, studied and travelled in more than fifty countries.*

TEA IN THE DESERT

Shirley Jee

Longlisted 2019 'Out of the Blue'

A bundle of coloured cloth was dropped into my lap, and a baby's eyes locked into mine. Smoke was taking its time rising from a small fire. The air was thick with the sweet smell of tea, poured from a blackened but elegant teapot. In the daytime gloom of the tent the Syrian Bedouin, their faces sculpted by weather and habit, were showing us one more of their most precious possessions.

First, we had met the shaggy brown goats outside bleating softly in their pens – a source of milk, yoghurt and cheese; then the huddles of sheep, also providing meat and an income when sold. Their woolly bodies and satisfied faces were strangely comforting in this barren desert landscape, with its occasional scattering of black plastic bags, worn-out tyres and discarded debris. And now, the youngest member of the family was being passed around for us to admire, like a special Christmas present. Half an hour earlier, our Syrian-speaking guide had instructed the driver to turn off the road. He had spotted a tent, a *buryuut hajar* (literally 'house of hair'), in the distance. Nicholas would respectfully ask these unsuspecting nomads if we could meet them. Our itinerary of ancient cities, citadels and archaeological treasures had just got even better.

Ahmad, the head of the family, was happy to show off a few gold teeth as he smiled his welcome to our group, sitting around the fire on layers of worn carpet. It was impolite to address questions to any other member of the family. 'How do you get to a dentist?' asked Rosemary, her gentle eyes and fine complexion preserved by a lifetime spent in libraries. 'When a white tent appears in the desert, the dentist is ready for business,' Nick translated. Agonising extractions came to mind. We learned later that Ahmad's resplendent teeth had only gold foil wrapped around them. After more questions, and fragrant tea sipped from tiny cups, one of the wives led the female guests through a heavy cloth partition. Richly patterned carpets, embroidered blankets for walls, a sagging cloth roof and heaps of coloured cushions created a kind of sanctuary. Women broke off their chatter and smiled self-consciously as we briefly shared their separate living space.

We left with a final farewell to the goats and sheep, indifferent to our studied reverence. But the Bedouin's love for their animals was palpable. We had already seen herdsmen, in their brown *tob* and red *keffiyeh*, pick out a sturdy-looking sheep and lift its two front legs on to their shoulders. Running their hands along the length of the animal's body, with the love and attention of a curator examining a Rodin sculpture, they were choosing the strongest animal to sacrifice at the end of Hajj: commemorating Abraham's sacrifice of his son, Isaac.

Two days later, in Deir ez-Zor, Bedouin women traded goods with the locals while their husbands walked untethered sheep along the street to the tiled shops for slaughter. Large wooden chopping blocks sat on tables outside. Behind them, men were busy chatting to old friends and sharpening knives. Every part of each animal would be carefully separated, to be used for food or clothing. We walked past wooden carts piled high with fleeces and innards; sheep's heads and hooves lay in separate heaps nearby. This orthodox community saw few tourists. I felt the stares. In our western clothes we were as alien to them as a bowl of sheep's eyeballs were to us – though this bond between human, animal and nature felt honourable. Only one young boy, perhaps a football fan, called out, 'Hello, England!'

Yet before leaving the Byzantine church of Mushabbak, sitting high above a village further north, two young girls had each held up a bunch of red anemones that they had picked for us. One dressed in a thick red tunic, the other in a stained white jumper, leggings and blue plastic boots; threads of dark straggly hair blown across deep brown eyes and shy faces. And when we pulled out of Damascus on the Hejaz Railway – steam billowing, carriages swaying and windows rattling as the train creaked and screeched its way out of the station – excited children ran alongside shouting '*Salaam!*' Older boys clung to the carriages for a free ride, smiling back at us with jubilant faces, before jumping back on to the road.

Sitting in the guardsman's seat, watching the desert recede as we travelled south through the fertile plains of the Hauran to Bosra, I wondered how we would have responded to strangers arriving at our home out of the blue. Would we have happily shared our tea, let strangers hold our babies or shown them where we slept? And would our children have ever picked wild flowers and offered them, smiling, to a stranger?

After managing a specialist archaeology bookshop in Bloomsbury, London, ***Shirley Jee*** *set up Museum Books from home, selling books worldwide for fourteen years. She has written for* The Guardian, The Times, Church Times *and* The Tablet *(as Shirley Lancaster) and remains a passionate traveller, with a particular love for Africa and its wildlife. Her trip to Syria is the one she most treasures.*

WHO ARE YOU?

Tom Swithenbank

Finalist 2021 'I'd Love to Go Back'

'Akto vi?!' The Lada had screeched to a halt in Ananevo's one of two streets and a square-jawed cropped-haired blonde male had jumped out and was now gesticulating at Robert and me as we stood, a little lost, a little bewildered, in the small Kirghiz settlement. We had arrived an hour earlier, staggering off the bus from Bishkek and looking forward to the promised resort on the shores of Lake Issyk-Kul that Ilya in Moscow had described in such glowing terms.

Promised resort no longer, it transpired. The rotten beams and the caving roofs of the wooden izbas by the lake shore suggested Ilya had not been here for a very long time, if ever. Dusk was setting in, we were six hours from the capital, we were not really supposed to be here; we had no visas – a concept no one had worried about too much as the Soviet Union disintegrated – and there we stood scratching our heads, wondering what next as the man strode ever more forcefully towards us. 'Akto vi?!' he repeated, a little louder now, as if we might not have heard the first time.

I looked at Robbie, I looked up and down the street and the lack of any definable escape route. Play straight, dissimulate? The lack of any entry stamp in my passport, either for here or Kazakhstan, still played on my mind. I took a deep breath. Play straight. 'Hi, I am Tom, this is Robbie, we were looking for the resort… ' I ventured in faltering Russian. 'Are you Latvian?!' he barked back. 'Your accent is

Latvian!' Another glance at Robbie, another deep breath. 'No, I am an Englishman, he is German.'

I could see a veil of confusion descend upon the man's face. Something clearly did not make sense. 'I don't believe you.' His stance had softened. His shoulders had dropped. The bulldog in him receded. The sun had now set behind the mountains behind the village, the high peaks shadowed against the purpling sky. Emboldened, I pulled out my passport and nodded to Rob that he should do the same. Holding firmly on to both I opened them to the photo pages. I pointed to 'United Kingdom' and 'Bundesrepublik Deutschland'. 'Videte? You see?' A moment of silence and a smile broke across his face. 'We have never had foreigners here,' he murmured. The significance of the statement for my recently liberated Latvian friends flitted through my mind. I put the passports back in my pocket.

'Come with me!' he ordered. 'Where do you want us to go?' Visions of police stations, lengthy discussions with officials, explanations of how two young students with a yen for adventure and an interest in the Great Game had decided one drunken evening to jump the very next morning on a train from St Petersburg to Almaty. 'Domoi!' Home. 'You must be hungry. You must meet my wife!'

Volodya led us the twenty yards from our point of hesitation through the gap in the painted picket fence, up the garden path to his wooden house. As we perched somewhat awkwardly on the divan, the obligatory carpet on the wall behind us, a bottle of *samogon* appeared as his wife peered warily at us from the kitchen. The night was making itself known through the window panes, a nervousness of where we would eventually sleep equally present.

We had been in Russia long enough to know the ritual. 'Za vstrechu!' I raised my glass and sank the undefinable liquid. My eyes watered, my throat burned. Rob spluttered. The glass refilled, 'To family!' As the clear contents of the bottle steadily receded, moonshine evaporating from our tongues as we worked through the roll-call of toasts, we sank deeper into the divan, the warmth of the room and the long journey behind us adding to heavy eyes. A polite enquiry about a hotel provoked hilarity and we were ushered into the front room where two beds had been freshly made. 'Hotel!' Volodya laughed, but one more for the road…

The night was long, friendships were sealed. Three weeks and a host of adventures later, with misty eyes and repeated bear hugs, we finally bid our goodbyes to Volodya and his family, our guilt at his refusal to accept any payment somewhat assuaged by the two hundred dollars tucked under the pillows. We corresponded for a while. The last we heard, Volodya was no longer a district nurse but, buoyed by an unexpected friendship and a small injection of capital, had become the village *biznesmen*. I would love to see him now. Kindness does indeed pay.

To a 3-year-old ***Tom Swithenbank****, to trundle your toy milk-cart two miles from your grandparents' to your uncle's house was self-evident, even if to him Nottingham was a foreign town in a foreign country. On safe arrival, the police search was called off. The thrust to be somewhere else burned bright and never waned. Motorcycles merely replaced the milk-cart.*

A GENEROUS SOLITUDE

Mairi McCurdy

Highly Commended 2020 'And That's When It Happened'

I balanced on the edge of the long drop. No – no Alpine cliff plummeted beneath me; no river snaked at the bottom of a gorge – I was in a cupboard-sized room with a broken lock at the end of a corridor in a straggle of buildings in Qinghai, China – the greatest danger: another guest barging in. The sensory assault of Chinese toilets was by now familiar, so with my head jammed against the door, I just closed my eyes and prayed it would soon be over.

We'd left the provincial capital, Xining, the previous day – a break to heal the ills of our suffocating work-unit in Shaanxi province, where partner Michel and I were teachers: all it would take was five hours... five hours and we'd be lost in peace and isolation in the infinite grasslands of the Tibetan Plateau, gazing across lonely Qinghai Hu: one of the world's largest saltwater lakes.

It took thirteen. Four of which were spent at Xining's Public Security Bureau after our bus driver performed a U-turn across four lanes of traffic. Then the 'development of the west' halted progress with an inconvenient road-widening scheme. Surrounding mountains were blasted to smithereens, showering the traffic with grit, while grinding engines hungered for the ensuing, honking gridlock. Finally, a few hours later, snowflakes slicing the headlights, we slid to standstill at a jack-knifed lorry. Our resourceful driver, Coke bottle of decanted diesel in hand, sparked the grounded vehicle to life and motored us onwards into the night. After thirteen hours of passive smoking, we staggered off the bus into impenetrable black.

The following morning, I was cramped in the smallest of rooms in a grimy guesthouse in Jangxiguo, salvaging some resilience for what next lay in wait. Gripping my pack, I stepped outside.

The snowstorm had gifted us azure skies and sharp, clear air. For miles, actually, hundreds of miles, the plateau stretched on and on – fringed with gleaming peaks and the endless steel of Qinghai Hu. Waves hugged the shore against ecstatic shrieks of gulls and geese. In front, behind and on all sides, there was… space… in incalculable quantities. Boundlessness unfolding all around, my ears strained for the hum of traffic or the clamour of a loudspeaker. But – no. This beauty was immaculate.

Our seclusion didn't last. Some boys tagged along, jumping on a dead seabird to show it could still squawk, then guiding us away from where the fierce dog lived; next a herder walked silently with us for a mile and as we cooked lunch, a local jumped from his horse to crouch and examine the mechanics of our stove.

Later, a 'plateau-pika' trapper, surrounded by hundreds of small desiccating mammals, explained how Chinese doctors would buy these mummies and boil them into a medicinal pulp to bandage on arthritic joints. Grasping a dead rodent by the feet, he demonstrated the flaying, peeling it like a banana. The fur was yanked off and discarded over his shoulder. With two mangy buzzards watching from a gatepost, he unzipped its belly and poked the tiny scarlet innards on to the grass. Ta da!

Lost in this treeless wilderness we hiked onwards, herds of yak parted in our wake, the stillness giving space for wonder at single sounds. And then it happened. A clutch of grey clouds clotted and swelled to consume the sky. The light hail, carried by the breeze, mutated into a vicious, swirling whiteout. Trudging blindly, we struggled on until a shape etched itself into view. Crouching in a dip was a small mud hut.

'Ni Hao!' Our shouts for attention were swallowed by the gales. In desperation I pulled back the tarpaulin. A woman appeared. Ushering us through the sculpted entrance into her two-roomed home, she threw sheepskins on the floor, inviting us to sit.

Window ledges and shelves, cut into the mud walls, had been smoothed, all sharp edges erased. Smoke from the mud stove curled up and out of a hole in the roof through which snowflakes now fell and sizzled in the embers. The woman shovelled handfuls of sheep droppings on to the fire. Thick hunks of fresh bread, bowls of tsampa and scalding butter tea were placed in our frozen hands.

Three generations lived here – none spoke Chinese. We shared biscuits while the men prodded our rucksacks, my pride on packing lightly shamed by the simplicity of their home. I felt humbled by the hospitality shown to us, these huge foreigners who might just demolish the fragile space with clumsy boots.

When the snowstorm passed, their broken-down motorbike presented us with a chance to repair and repay. Offering money would offend, but perhaps we should have 'lost' some yuan behind that sheepskin.

Walking to the edge of their horizon, our new friends still waving, we were nothing but pinpricks on their generous solitude.

Mairi McCurdy *teaches and writes in Newcastle, County Down. When not educating 21st-century teens, she'll be found hiking in the magnificent Mourne Mountains with her family and grabbing her suitcase as soon as* *school's out. She has lived and worked in Europe and Asia, had her writing published by Bradt and is currently writing a book based on her two years of development work in western China. She has also just completed a novel set in Germany at the time of the fall of the Berlin Wall.*

4
FEATHERS, FINS & FUR

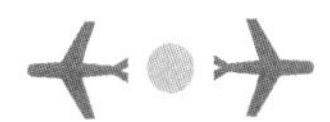

"He can't see us, he can't smell us and he isn't expecting us. He is in his world and we are in his world."
Lynn Watkins

Our tracks and trails are often crossed by creatures. Whether carefully planned and longed-for, or serenely serendipitous, these stories record those breathless, unforgettable moments.

**Nepal France Sudanese coast
Alaska Cameroon Zimbabwe Mozambique
Sri Lanka Malaysia Mongolia Borneo Yukon**

THE TIGER'S TAIL

Dom Tulett

Winner 2016 'A Brief Encounter'

A swollen sun, blunted and smudged by morning mists, watched with me as elephants emerged from the forest. They moved slowly, gracefully, and when their trunks swung wide enough to the side I was sure I could see them smile. I was reminded of the previous evening; a boy from the village approached me as I sat with a beer watching the day fade into the river. He handed me a slip of paper. On it was written: *NEPAL. Never Ending Peace And Love.* 'I hope you like my country,' he said, and he smiled and walked on along the riverbank.

The morning was cold and wet. A dew glossed every surface and the air clasped my throat and chilled my breath. I fought to puff out clouds of white. The elephants were our transport; our ride into the Chitwan National Park. Gopal, alert and lean, forever blowing on his fingers to keep them warm, was our guide. He had been doing this for twenty-two years. He briefed us on what to expect, and what not to expect: tigers landed in the latter category; squirrels in the former; rhinos somewhere in the middle.

I was sat next to a fidgeting Norwegian girl. It was her first time trying to see a tiger in the wild, but I had been here many times before, and always failed: India – nothing; Malaysia – nothing; Bangladesh – a paw print in the mud. She asked me if I thought we would see a tiger.

'I hope so.'

'I think we will,' she said, repositioning herself on the creaking sedan chair. 'I'm a bit lucky like that.'

The mist had begun to burn off as our elephant took its first steps towards the forest. We moved clumsily, with a heavy lateral swagger – roll and pop – that jolted right through me on every other step. The sun gained colour and strength as we followed a muddy trail rutted with long puddles, and it reflected up beneath us and rippled out with the elephant's strides. A range of greens approached: a patchy, pale tint of the tufted parkland; a lusher wave in the dense tall grasses; and a darker block where the trees met the sky.

Indonesia – nothing.

We broke through the treeline and almost immediately Gopal stopped us. His eyes narrowed and his head tilted. He blinked rapidly three or four times. I tried to follow his stare into the undergrowth. I looked but saw nothing, only clumps of vegetation and patterns of leaves.

Gopal blew on his fingers.

Deep greens and thick browns.

I tried to breathe silently, but the harder I fought, the louder I could hear myself. My blood thumped through my veins, pulsing in my ears like a bass drum.

Pipes of sunlight and hollows of shade.

Then Gopal sat straight upright, stretched his back and unzipped his jacket. He blew on his fingers and, with a look of complete indifference, we pressed forward again.

Thailand – the remains of a kill.

Our elephant walked on and on – roll and pop, roll and pop – for three hours more as we roamed wide through the forest. There were many false alarms; each time we paused, strained our eyes, resigned, and moved on. Throughout, the forest hummed with noise – chirping and chattering. Gopal pointed out other wildlife: a sulking boar, an angry woodpecker, neurotic deer. The Norwegian girl had stopped fidgeting. She whispered in frustration, 'We have better deer than that at home.'

India again – nothing.

The forest fell silent. The elephant paused. Gopal held out an arm, dipped his neck and bore his eyes into the layers of tangled foliage. I watched him closely. He did not blink, did not blow on his fingers. Nothing about him moved, except for his mouth, which breathed a longed-for whisper, 'There.'

My heart locked. 'Where?'

'There. Past the silver tree with the split trunk.'

Again I looked but saw nothing. I leaned forward for an inch of improved view. The sedan creaked and scared a bird from a nearby branch, stealing my attention and gaze for a moment. When I looked back down, I saw it; hovering in the smallest gap in the leaves, almost out of sight, yet I've never seen anything more clearly. It was there. Rings of white-orange and black, looped upward at the end in a curve, like a smile. The entirety of the forest reduced to a gap in the leaves. A lifetime's dream filtered into one moment. The tiger's tail. It was there.

Then the tail twitched and flicked, and slid out of sight again. And that was it; that was all I saw. Nepal – something; for just a moment, a pulse of the universe – no longer than a breath. Elusive and beautiful. There and gone.

***Dom Tulett** started entering travel-writing competitions after the birth of his daughter, as opportunities for travel went the same way as sleep. He wrote to remember. Following his Bradt win, Dom also won the National Geographic Traveller annual travel-writing competition, had pieces featured in Bradt's travel-writing anthologies, and has written for* The Guardian *and* The Telegraph. *His sleep has since improved.*

A WOLF IN THE MOUNTAINS

Julia Bohanna

Winner 2012 'A Close Encounter'

My grandmother is talking about Lourdes' miracles, even though the town – with the light up Virgin Marys and rabble of rosaries – is far below us. Please stop turning around, Nana. I am behind her, on a donkey called Pierre, weaving through the Pyrenees. The donkey has sides like an overstuffed purse, a trembling mouth and the dark hopeless eyes of a depressive. A caravan of us making clok-clok on a stony mountain path is following an overconfident Gallic guide who is now far out in front. The most eager riders are looking for golden eagles or perhaps a spiky-eared scops owl left over from night hunting. Mist laces the mountains. It is too cold and too early to be truly appreciative. I don't want birds and I don't want miracles.

I am being punished for the sin of setting fire to my cardboard candleholder at Midnight Mass the night before. Hooded nuns around us hissed disapproval like cobras, as fire made chaos in a crowd. I was trouble to them – uninterested in Bernadette, the grotto or the collective healing faith of the place. I didn't feel it. I was happy in cynicism. There as companion only.

'Isn't it beautiful?'

Nana's reedy voice is sucked away by the mountains.

'There are bears in the Pyrenees you know, Nana. Huge claws and hunger instead of humanity.'

There is no provoking her but at least the stories of miracles and hope are fading. Her donkey moves so far ahead I finally cannot hear. Tomorrow we visit Bernadette's waxy preserved body in Nevers. More morbid devotion.

My donkey has stopped. He sighs. In a strange hoofy tiptoe, he moves – not forward – but sideways towards the edge. Stones flake from the side and give me an indication of his wish to die.

I think we both see the eyes at the same time. In the trees a few feet away, watching us. Not human eyes. They are amber. An animal. A large animal. The donkey's head jerks up and he makes a sound, the controlled squeal a child makes when hiding, afraid. Suddenly, filling plastic bottles of the Virgin Mary with holy water while an old devout woman waits doesn't seem so dull. I crave the thin soup in the hotel or the cool balcony where I can watch the mountains. Or even the wheelchair conmen who beg and pack up at night to stride home.

Stones are still falling into the valley. The mist is clearing a little. Maybe it is one of the huge Pyrenean dogs out there. I listen for goat bells or the ma-ma bleat of a young mouflon lamb, which I know graze here. Or better still, a reassuringly smelly shepherd wearing a neckerchief.

There is nothing but those strange, steady, mesmeric eyes.

Now I can see a head, then a body. It is doglike, crouching. They are the long muscular legs of a large, black-brown wolf – but with a grey muzzle. Old. Slow. I would like to see Lourdes again, not to be picked off as the weakest of the herd. That's what wolves do… they separate the weak and the sick, surround them and tear off strips of flesh like liquorice. Don't they?

There is something that holds me in the wolf's gaze: a brightness, a soulful connection of sorts, despite the fear. We stare one another down – one species to another. Canis lupus. My donkey's chocolate flesh is now dewed with sweat. He ignores at first entreaties from my legs for him to move. But finally, with me still straining to keep my eyes on the wolf, he moves on, much faster and with more purpose than I have seen him before. Still sweating but now comically pretending youth and vigour. Head erect. Every purposeful hoof-clop stakes its claim and says: I am not prey. The wolf stands for a moment like a fond lover, then turns, tail down – back down into the valley.

Later, our guide in angry spittled French will deny the wolf with a wave of a ringed hand. There are no wolves in the Pyrenees, he says. He would never put people at risk. Years later, I will read that they have indeed mysteriously returned to some areas and that farmers are taking measures to protect their herds. More dogs, red flags called fladry. Measures.

I will already be working in wolf conservation. I have put my fingers through a wolf's belly fur, cradled cubs, walked alongside them and been caught in the gaze of those mythical eyes. Fallen in love with a species even when I am giving them fly-raddled sheep's guts from a bucket. There is less cynicism in my life, replaced by passion.

My grandmother talked about miracles for a long time afterwards, but I brushed them aside.

This atmospheric piece won the Bradt Travel-writing Competition in 2012. Since then, we have lost contact with the author.

SHARING THE SKY

Paul Alexander

Longlisted 2012 'A Close Encounter'

Sacred peaks float above the haze away to the north. Below me the slope of the hillside drops down to terraced fields. My eyes follow the contours of the ridge and I spot something circling below. Pulling on a line, I lean towards it, gliding in an arc towards the distant bird that is soaring in wide circles. We drift towards each other. I stare down at the tiny brown wings that rise towards my own.

Then my paraglider bites into the thermal. I lean more, banking tighter. The earth starts to turn beneath me as I try to focus on staying in the bubble of rising air and the forces of the universe take me upwards. Still turning, then suddenly the griffon vulture is in front of me – its pale head and yellow beak focussed on the ground in its search for the scent of carrion. Suddenly its wings seem the size of a house door and our paths are converging, fast. I freeze, fixated by the approaching danger, then shout – 'Oy!'

The vulture folds its wings, a giant black W hisses below my feet. I curse, and watch it glide away along the slope, to land on a crag.

Later, I am in the landing field with Sandip and Madhav and Ashok, watching the same vulture soar again a thousand feet above us. We heave our packs on and walk along a lane of dark pink dust, between fields of maize and wheat. Wood smoke drifts upwards beside stockades and chickens peck the ground. Past homes of mud brick, where bright flags hang and the air is still, and silent, except for the chip of grasshoppers and the shuffling hooves of cows that watch us pass.

My companions jog up the path. I plod behind. Going up is hard work now. I try to focus on where to place my walking boots while slim, bow-legged Nepalis skitter down the rocky path in sandals.

Finally, we reach the village of Bandipur by the flagstone steps of a path that runs through the orchard behind my guesthouse. The street is lined with three- and four-storey houses of stone and earth. Beams prop up long balconies and sloping porches. Below flaking wooden shutters, hibiscus and geraniums are in bloom.

We drink cold beer. A woman makes dumplings in a wok in the shadows of a kitchen. Her daughter sits combing her hair in the sunshine.

'Buy me something?' she smiles. 'Buy me a fried egg?' And five minutes later she is back, grinning, with dried yolk in the corner of her mouth.

Later, it is our turn to eat – the ubiquitous daal and rice. The local pilots watch me eat, and my approval is met with grins. Lights go out and in the power cut they order rum. Oil lamps burn, showing nothing more than flickering faces with eyes closed as they break into ballads. The choruses are loud and joyous, defying the silence of the mountains, the darkness, and the chill of the night. Three passing backpackers are invited to join in.

When the party is over I make my way up to the guesthouse, lift a heavy brass knocker, and my final climb of the day is up narrow, wooden steps to frugal silence.

Hot tea, and morning sunshine. Fingertips brush flecks of yellow and red paint on the doorways as a woman makes her way from house to house. Children ring single chimes on a heavy bell as they pass, in their blue school uniforms, while older figures cloaked in crimson spin golden prayer wheels on the wall of an adjacent shrine.

'Where to, next?' I ask the backpackers.

'The temple of Manakamana. And you? Are you going to fly again today?'

I look up at the crest of the hill above the village, remembering my encounter of the previous day. I want to say, 'No, just watch. And learn.'

But I just nod. 'Yes, I may as well make the most of it while I'm here.'

***Paul Alexander** is a retired language teacher who attempts to string together sentences that occasionally resemble travel writing, flash fiction and short stories. His work has been published online (Inkapture, Ether, Jottify), placed in a competition anthology (*Storgy *vol. 2), read at spoken-word events (White Rabbit), seized by HM Customs, and lost in the post. He lives near Stroud.*

SEARCHING FOR MERMAIDS

Liz Cleere

Finalist 2011 'Up the Creek'

As the reef rushes towards me, I grip the slippery stainless-steel wheel, my hands sweating and prickling with fear. The sea moves from lapis lazuli to sapphire, then turquoise, emerald and jade, finally fracturing into a myriad piercing colours. I've no time to admire its beauty; I must hold my course. Jamie shouts words of encouragement. 'For God's sake, concentrate!'

For millennia ancient mariners have told of mermaids, with their mesmerising beauty and seductive songs, luring sailors to their death on sharp-toothed rocks. Today, as we sail our thin-skinned boat through shards of coral, I'm beginning to understand where those old Jack Tars were coming from.

Dolphins peel off into the navy blue sea, and a hitch-hiking tern flies away. Balancing high up on the boom, Jamie peers into the now crystal water, navigating by sight – and the seat of his pants.

Dugongs gave rise to the legend of mermaids. One had lived on my desktop at work, but I left it behind when I exchanged the daily grind for a life at sea. Distantly related to elephants, and looking like a cross between a walrus and a potato, these gentle vegetarians – they eat nothing but seagrass – grow to three metres in length and weigh half a tonne. When a fellow sailor whispered to me that she had spotted one in a hidden marsa on the Sudanese coast,

I persuaded Jamie to stop there on our southbound passage through the Red Sea.

'Five degrees to port. Don't lose your speed.'

Marsas are narrow winding creeks, the desert equivalent of a fjord. We found this one easily, but manoeuvring the boat – our home – from the safety of the deep Red Sea, through a kilometre of crooked coral-fringed channels, is proving to be even more of a challenge than we'd anticipated. I follow Jamie's command and shrug off the instinct to slow down – I mustn't lose momentum.

Our cat, oblivious to the danger she's in, leans overboard, ears twitching as her X-ray eyes follow hundreds of startled fish beneath the bow. She is not alone in savouring the smorgasbord laid out in front of her: an osprey, with the speed and precision of a missile, swoops down to pick up its dinner.

We glide round another hairpin bend and the marsa bursts open. A welcoming lagoon, ringed by a creamy golden shore, folds us into its protective arms: we've made it up the creek, paddles intact. After dropping the hook on a sandy seabed, with a healthy carpet of seagrass growing on it, Jamie gently peels my clenched fingers from the steering wheel. We've stumbled into a David Attenborough BBC documentary: African savannah stretches to distant mountains in the west; to the south the sea tickles a pristine beach, and on the northern shore ten-metre-high fossil-filled rocks jut over the water. There's not a human in sight. Neither is there a dugong.

On cue, the late afternoon desert wind sweeps in, whipping up the waves. I scan the surface for a fluked tail. Nothing. Unperturbed by a 44-foot hull appearing in the middle of its afternoon perambulations, a turtle silently swims past. For a millisecond I mistake it for a dugong and am ashamed at my disappointment.

When dusk arrives the wind drops, smoothing and quietening the lagoon. But not for long. Soon the water is boiling with fish: the familiar sound of nocturnal behemoths herding their prey. After our own supper of tuna, caught off the back of the boat earlier, we relax in the warm night air. Later, while Scorpius crawls across an ebony sky, we fall asleep.

Dugongs share a need for sun, sea and sand with package-holiday makers, but they are shy creatures, and are losing the battle to jet skis and sunscreen. Designated 'vulnerable' on the IUCN list of endangered species, they are victims of our human population explosion.

Over the next few days, we explore the marsa: sting rays glide in the shallows; a Goliath heron stands poised for the kill, and hermit crabs (in a variety of recycled homes) play hide-and-seek with me on the beach. I collect bits of jetsam, driftwood, broken coral and two football-sized, empty conchs. Jamie finds the osprey's nest, complete with chicks, tucked away on a rocky outcrop. There are no dugongs.

A few days later it is time to leave our haven: the wind has turned southerly and we have an ocean to cross. Reluctantly I take the wheel, while Jamie lifts the anchor. Then a dugong comes up for breath, and time stands still. I am enchanted. It dives and resurfaces, trawling for food, a graceful and stoical, giant spud.

As we drift towards the rocks, a shout breaks the spell.

'Say goodbye to the mermaid.'

Liz Cleere *and Jamie began their slow passage to the east through mermaids and pirates in 2006. Enchanted by southeast Asia since 2013, their favourite travel is in remote places like the Hinako Islands, Anambas archipelago and Balambangan Island. They are regular contributors to* Yachting Monthly *magazine and blog about sailing and travel at followtheboat.com, and Followtheboat*

ALASKAN ADRENALINE GRASS

Lynn Watkins

Commended 2016 'A Brief Encounter'

I never thought I'd be so pleased to see someone holding a gun. It's held firmly upright, by Patrick, our fishing guide, who is kneeling behind us on the edge of a gravel bed in the middle of the Alagnak River in Katmai, Alaska.

We have been watching a lone grey wolf, trotting slowly along the opposite river bank. A rare daytime wildlife treat, we catch glimpses through the grasses and tundra shrub that surround both him and us. Tempted by the thought of an even better photo, we have abandoned our fishing rods and boat, and crawled our way to the edge of the gravel bed. The wolf, patrolling up and down the edge of the cool water, seems unaware of us and we savour the wilderness performance.

We are similarly unaware of a brown bear further upstream in a meander in the river. He is snorkelling towards us, his face submerged and the water up to his ears. He is intent on the sockeye salmon that are going to add to his thermal layers to survive winter. He can't see us, he can't smell us and he isn't expecting us. He is in his world and we are in his. I'm still hunkered in the grasses, absorbed, watching the wolf until I notice a rustling. Patrick mouths the word 'bear' and 'ssshh'. He has the gun in his hand.

Our boat is bobbing gently in the current, visible, taunting, but a foolish distance away.

Patrick whispers something. I catch only one word. Run. 'Never surprise a bear and never run,' echoes Charlie's voice from our earlier safety briefing at the fishing lodge. I can feel myself trembling. We must really be in a predicament if Patrick's suggesting running for it.

I want to run. Badly. I glance at the boat. It's impossible of course, not least because of my own constricting thermal layers. All the night-time reading I've done about bears flashes into my mind. Even female bears average about 300lbs and males can weigh as much as 1,000lbs at the summer's end. It's a good idea to make a lot of noise when walking in bear country to avoid any startling encounters. Hiding in the grass they are heading towards wasn't featured.

'Run?' I ask. 'No, never run,' is the calm, firm answer. 'Stay crouching. When I say so, we'll all stand up together. Slowly. We'll look bigger grouped together.'

I can't see the bear but I can hear him. Despite the whooshing pounding of my heart in my ears. The quiet ripples of the river are disturbed by splashes and the droplets of spume. He's nearer now. I feel a hand on my shoulder and I turn to see Patrick pointing to us and then up.

The bear lifts his head up out of the water and we all rise together.

I don't know whether a bear has eyebrows but if he does I'm sure they hit the top of his forehead. He soars up, standing on his back legs, water dripping from his dense fur. I'm looking directly at a huge brown bear, darkened almost black by the water. Exactly the sort of photo I would like to take. There is little sound, just the distant piercing call of a sea eagle and the soothing voice of Patrick repeating 'hey bear'. The gun is pointing skyward, ready for a warning shot. The bear continues to stand still in the river, watching, calculating and then with a sudden twist, turns to his left and bounds off to the opposite river bank.

Safely back in the boat, and flask of coffee opened, we breathe more normally again as we slide into the currents of the river.

'Did you get a picture of the wolf?' asks Patrick.

'No. There was too much grass in the way.'

As a researcher and policy adviser, **Lynn Watkins** *has done a lot of writing – think huge reports and boring footnotes. There is no room for personal opinion, emotion or feelings. Having entertained friends over dinner with travel stories (they will laugh at anything for free food and wine), she thought she'd have a go at writing one down.*

THE GREAT ESCAPE

Grant Hackleton

Finalist 2007 'A Chance Encounter'

4.29pm

'Custard cream, anyone?'

There was silence, then a few uneasy giggles, then silence again, except for our nervous breathing. It was a strange question given our predicament.

There were six of us, white tourists, in a tree, in the dense jungle, at the base of Mt Cameroon, West Africa.

'Do you think we're safe up here?'

'I think so.'

'How tall can a forest elephant be?'

'They can push trees over.'

'This is a pretty big tree.'

'My shirt is torn.'

'Yeah, it is. And you're bleeding on your back.'

'I can't even feel it.'

'I've grazed my knee. I didn't feel that.'

'At least we're all safe.'

'I've never climbed a tree before.'

'You learnt.'

'You're a very fast runner, Bill.'

'I know. I surprised myself. I didn't want to be last.'

'When I ran past you Laura, you were crying. How can you run and cry? Run first, cry later.'

'I was so scared.'

'So was I.'

'I'm still shaking.'

'Me too.'

'Are we going to get down?'

'No!'

'How long do we stay up here?'

'All night. I don't care. I'm not getting down until they've gone.'

'How will we know that?'

'I don't know.'

'Pim tripped up.'

'I know. I saw him.'

'I'm sorry that I didn't stop, Pim. I was so scared I was really running for my life. I'm sorry I left you behind.'

'I stopped.'

'You're married to him.'

'Someone helped me up the tree. Someone pushed my foot up when I was climbing. I couldn't get a grip and then suddenly someone helped me. I didn't even look back to see who it was. I just trod on them.'

'That was me.'

'I'm sorry. And thanks.'

'I'm serious. Does anybody want a custard cream: I've been carrying them for the last four days.'

'I'm not hungry.'

'Me neither.'

'Where are our guides?'

'They all ran the other way.'

'Abandoned us.'

'Saved themselves.'

'Do you think they've gone?'

'The guides?'

'The elephants.'

'Why don't you get down and see, Laura?'

'Yeah, right.'

5.14pm

'The noise was incredible. I have never heard such loud noise.'

'When the bull trumpeted?'

'Three times!'

'All of it.'

'One minute we're in a forest. It's quiet and the birds are chirping. We're chatting away, saying maybe we'll see one of these rare, elusive, forest elephants, maybe we'll have this amazing African wildlife experience, then all of a sudden, chaos! Whole saplings were coming down around us. I could hear branches snapping. The canopy was falling on us. The ground was shaking. The bull elephants trumpeting at us. We're running like hell. Our guides have gone. It was like a train had derailed and was crashing through the forest. We were so lucky. I'm serious. The noise was incredible.'

'You've got to love Africa.'

'The worst thing was not knowing where they were coming from. Hearing them crashing through the bush but not being able to see them. It's so thick it's impossible to see. It was like running blindfolded. Wondering whether we were running towards them, or away from them. Not knowing where they were. I didn't know whether I was running into the herd or away from it.'

'I didn't think that much about it. I just ran.'

'We were lucky, though, hey?'

'Yeah, lucky not to get killed.'

'No, lucky to interact with such excellent creatures.'

'Interact? They charged us.'

'It was great.'

'They tried to kill us.'

'They didn't.'

'If we hadn't run, we would have been killed.'

'It's instinct. They were just protecting themselves.'

'That was the worst experience of my life.'

'Or the best. How many people have discovered forest elephants? And heard one bellow! Sam has been a trapper in the forest for six years and he's never even seen one. This was a once-in-a-lifetime opportunity.'

'Are you joking? We were so stupid to get so close.'

'We didn't know they were there.'

'We knew they were somewhere.'

'Yeah, but not right there. Close enough to charge us.'

'We were stupid.'

'It was wonderful. Hardly anyone has seen a forest elephant. It was great.'

'You're mad.'

5.17pm

'I didn't see them, you know. I really didn't. I was too busy running away.'

'Well yeah. I heard them. Damn, I heard them, But I didn't see them either. I didn't look back.'

'Nor me.'

'Did you, Monty?'

'No.'

'Did anyone?'

'No.'

'You're joking? None of us saw them. Not one? Damn that's unlucky.'

'We were lucky not to die.'

'We were so unlucky not to see them.'

This ambitious and original story was a finalist in the Bradt Travel-writing Competition in 2007. We have since lost contact with the author.

THE ROAD

Angela Moore

Finalist 2007 'A Chance Encounter'

Turn left at Mukuti and the road gets exciting. It twists and rolls down into the Zambezi Valley, into the real bush, towards the setting sun and home. You can start watching for wildlife now. Battered, lurching road signs promise elephant crossing and kudu leaping. Graffiti, poignantly faded, promises Mugabe will go.

The road is mapped with family mythology. A hundred stories: the valley where we always see buffalo. The two sentinel baobab trees. The place where we ran out of fuel. The tree my brother swore blind was a rhino. The first view of the lake. The map's a little faded for me now. I'm not as certain of my way as I once was. I've been ten years away from home; the country's changed so much. It looks poor, hungry; a desperate Third World face for my halfway First World eyes. Coming home, I feel twice a foreigner.

This is the road we're driving now, heading for a weekend at the river. My parents, greying with the exhausting grind of living in Zimbabwe, sit in the front of the Land Rover. Kie and I perch erratically in the back, amid mounds of canvas and cooler bags and crates of drinks.

The winter bush bumps past us in a hundred lovely shades of dry. We've not seen much game. It's too late, too hot. The blazing time of day when only the chaotically late to leave are anywhere but the deepest shade bush.

I'm hanging on, battling with the wind-whipped tangle of sarongs I have wrapped around me against the sun, when it happens. A flash of dusty flank, yellow eyes, flicking tail. They're so close that I could lean down and touch them. The shock of it locks my tongue for a second, then I'm banging on the window, boggle-eyed, shouting, 'Lion!'

My father slows the Landie to a crawl, yells back at me, 'Lion? Really?'

'Lion! Just back there! Just off the road.'

I'm incoherent, excited by this unbelievable stroke of luck.

'OK, we'll go back and take a look. Say when you spot them.' 'Watch it though, they're right on the road – really close. Seriously close, not tourist close. Kie, watch it, they're on your side. Hang on. Careful,' I jabber.

Then we're bumping slowly back along the road, the engine suddenly loud. I can smell the bush. The heat shimmers white on the tar. Further and further back, and I just can't see them. Still can't see them, still can't see them, getting panicky, where the hell are they! The tension's rising, we're slowing and slowing and then – almost unbelievably a row breaks out in the front of the Landie.

'Stop! Stop, you silly bugger,' shrieks my mother, who has spotted the lions.

'Calm down! Calm down, for God's sake,' bellows Dad, who hasn't.

'You'll get them eaten! It'll eat the kids!' yells Mum.

'Where are they? Where are they?' I whisper-shout, dry-mouthed.

Matters are settled – unequivocally – by an explosion of sound and movement, some twenty feet ahead. Enraged by our dithering, the lion goes from lazy tail-flicking to full-blown furious charge. It's so fast that not one of us sees quite how it happens.

I don't register the jolt as my father throws the Landie into reverse, or the screaming of the engine as we shoot backwards up the road. I don't hear Kie cursing or check if he's still on board. All I see is the burning intent in those wrathful yellow eyes as the king of predators comes belting down towards me.

I hear the grunt as he gives up the chase. The Landie rumbles to a stop in the sudden ringing silence. It's that moment. The white-knuckled heart-racing suspended moment, just before the brain and the talking and the adrenaline giggles kick in. I catch a glimpse of angry cat, tail ramrod-straight, heading into deeper bush.

'Want to go back?' yells Dad, hoarsely.

'What??'

'Want to take another look? Kie's first big cat in Africa?'

'No! That's fine!'

'You sure?'

'Really sure. We've seen him. Really.'

Later, we told the story around the fire, in the all-enveloping velvet of the African night. We laughed, argued over what we'd seen.

'That black mane…'

'Dinner-plate paws…'

'Landie didn't pop out of gear…'

'Didn't fall off the back…'

'The only thing moving faster than inflation in this country…'

And that night, as I lay waiting for sleep, I could still catch faint traces of the moment. That bright, clear moment, when everything – home, family, leaving, Zimbabwe – everything fell away.

This poignant and evocative story was a finalist in the Bradt Travel-writing Competition in 2007. We have since lost contact with the author.

A SMALL WORLD

Tom Franklin

Longlisted 2012 'A Close Encounter'

What we do is travel with bicycles. Mostly we ride them, sometimes we push. I could say that we are round-the-world cyclists but you might place us upon some high pedestal of fitness and determination on which we don't belong. Really we are cycle tramps who, with no great plan, bumble along the paths and back roads that link together villages and towns, countries and continents. We don't see so many sights, we live among everyday things.

I could of course write about close encounters with armed nomads, of Indian elephants and rabid dogs, of crocodiles and snakes, of Asian city traffic or road trains in the red Outback. I could do it in that typical English way; mentioning but playing down any danger or courageousness; laughing at our own naivety when confronted with outlandish ways, but leaving you with the impression that we are really rather adventurous to have ventured out into such a strange and hostile world. The truth however is that though it is often strange, it is very seldom hostile.

Along the way we have seen of course many birds and beasties and met so many people, some of whom we can now call friends but our constant companions, the ones who are always there building and destroying, eating and scurrying, creeping and crawling, reproducing and dying before our eyes are the ever-present insects. The encounters closest to our hearts are with this curious and wonderful world beneath our feet and our most numerous and memorable moments have been with six-legged species.

The passion began with the ants. With the exception of New Zealand where strangely not much is afoot, ants are pretty much everywhere. Each time we stop to brew up tea they appear to clean up our crumbs. We try to discover an order in their actions; six ants gather round a piece of biscuit and all pull in different directions, left a bit right a bit, like removal men with a piano they eventually reach their goal. Slowly the crumbs are lifted and dragged towards a small hole in the ground; those that don't fit are cut up and within twenty minutes all is tidy once more. I hope the high sugar diet doesn't cause an outbreak of obesity.

The smallest creatures so far encountered were some tiny round mites who were bothering a handsome French beetle. A half-hour operation removed all five of them and the beetle only lost one leg in the process. I am sure he limped happily away. The longest yet met was a six-inch praying mantis that thought my bare leg was a tree but the largest by far was a giant moth we pried from the beak of a kookaburra in Australia (well, he dropped it when I threw a wayward stone at him).

My personal favourites were the giant Indian millipedes who trundled along the paths like country trains; seemingly indifferent to being picked up, they wound their way around our fingers, tickled us with their many tiny feet then carried on along their invisible rails.

I am certain we have discovered insect species yet undocumented by man. Some creatures so small and insignificant they must have evaded study and classification. I hope they can avoid pests, pesticides and extinction. There is still a world here to explore.

One day on a woody lane somewhere in southern China where water seeped from a clayey bank and moistened the summer dust, a million white feathers came for a day to drink and to dance. The road was hardly visible through the bustle of half-inch, plumed hoppers strutting their impressive stuff. We had seen not one before this fifty-yard patch and never saw one again. Our research revealed no creature vaguely alike until, in a damp village hall near Christchurch a thirty-year-old *National Geographic* fell open at a photograph of a similar African bug which, when gathered in large numbers, cannily resembles a particular flower and thereby avoids being eaten.

Of course I could write about mosquitoes and flies and of bedbugs and spiders and encounter them we do; they cause more an occasional discomfort than a constant problem. We have far more experience of beautiful, inscrutable insects and though we are far from being experts or entomologists, as travellers, insects are a big thing in our lives.

Whenever you find yourselves in a foreign land and have visited the temples, done the rounds and got bored with the bus, sit for a while and see what wanders by, lie on the ground, climb a tree, scrabble in the dirt. Get close to where it's really going on.

Tom Franklin *Devon child with goats. Teenage tramp with dog. Archaeologist in tent. Chisel in hand on cathedral. Hammering iron in Dorset forge. Surrounded by kids in an old French house. Bauhaus building in German forest. Frying sausages by a roadside. Somewhere on a bicycle. Somewhere in a canoe.*

WHALE SHARK CENTRAL

Henry Wismayer

Longlisted 2012 'A Close Encounter'

'People might look for these fish all over the world – Thailand, Dominican Republic – see nada. Then they come to Whale Shark Alley and they can't believe it.' Isham squints towards the ocean with the narrowing eyes of a big-game hunter. 'The sharks are out there, alright. There's an 80 per cent chance you'll see one today.'

It was always going to take something special to help me overcome the do-nothing imperative of Praia do Tofo. Cutting a golden arc along the central Mozambican shoreline, 250 miles up-coast from Maputo, this is a beautiful beach calibrated to African time, where life slows down to a sun-soaked loll of wave-watching, seafood grills and naps in the shade of palm-frond cabanas.

Yet there is an overarching incentive to get off the sun lounger. For the last few days, I'd been hearing about little else: from the beach-boys looking for kick-backs to the South African drifters downing shots of Tipo Tinto rum in the bars that line the dunes, word on the sand was: 'The sharks are out there...'

Ordinarily, this refrain would be a good reason to stay the hell away from the ocean, but Tofo's sharks were different. These were whale sharks, the gummy goliaths of the shark family, and rumour had it there was a veritable posse of them cruising just offshore.

The idea of swimming with the ocean's biggest fish is not unique to Tofo. Although the whale shark is endangered, its wide dispersal makes this one unforgettable wildlife encounter that can be sought throughout the world's tropical seas. But there is nowhere else on earth where your chances of finding one are quite this good. A recent scientific study found that excursions out of Tofo ran an 87 per cent chance of encountering the big one. Stump up US$40-50 to one of the handful of dive shops that run these daily 'ocean safaris' and you could strike it as lucky as the group that saw fourteen whale sharks in one two-hour trip, or the blessed bunch that swam alongside one insouciant monster for 49 minutes. Isham, the old-hand who'd just sent me down the beach with his gravelly 80 per cent reassurances ringing in my ears, had undersold it.

Half an hour later, I'm barrelling over the tumble-dryer surf in a supercharged rigid inflatable alongside fifteen other tourists, hoping that Tofo lives up to form. Trying to cater for our high expectations are the Mozambican crew: a grizzled mariner at the wheel; a spotter, perched on top of a triangular scaffold above the stern to spy for fins; and Rafael, our diminutive guide, already rattling out the etiquette we would need to follow to ensure that the fish will be only mildly inconvenienced by our intrusion: 'Don't swim in front of it, don't swim by the tail. And whatever you do, don't touch!'

Soon we arrive at Whale Shark Alley. This is where the sea is at its most nutritious: an invisible spaghetti junction of plankton blooms, the filter-feeder equivalent of an all-you-can-eat buffet. The pilot kills the motor a few hundred metres offshore from Praia da Roscha, where a trio of surfers are waiting for the perfect break.

For a while, it seems like they have struck upon a much better way to exploit the conditions. On calm days, when the sea is flat as polished topaz, spotting marine mega-fauna is easy. Today, tossed about in a constantly rippling desert of water, it seems a lost cause. A passing school of flying fish provide an iridescent injection of excitement. A girl at the back pukes over the side.

I'm just preparing my appeal for a refund when there is a flurry of pointing and an urgent Portuguese exchange between Rafael and his amigo on the scaffold perch.

'Put your masks on everyone.'

'What is it?' I ask, and then see for myself – a grey fin cutting low and smooth, twenty metres off our port-side: a dolphin?

Rafael and the pilot turn and say in unison: 'Whale shark!'

It is as if a grenade has just landed in the boat. Within seconds, sixteen backsides have swivelled on the rubber; sixty-four limbs have flailed through the air and sixteen flippered tourists have plopped into the Indian Ocean.

The sight that greets me as the water closes in overhead is akin to the last image Jonah saw the moment he entered into biblical lore: an oval mouth big enough to swallow me without touching the sides. Its owner is a giant, a mature male whale shark, over eight metres long and weighing as much as six bull elephants.

And slowly, as the motor-drone fades into the background and everyone chisels out some doggy-paddling territory where they won't get inadvertently swallowed, the shark's hypnotic tail-fin begins to cast a magic spell.

The sharks are out here, alright. And I've just swum with one.

Henry Wismayer *is a freelance writer based in South London. His essays, features and commentary have appeared in over eighty publications, including the* New York Times Magazine, The Atlantic, Washington Post, *and* TIME Magazine. *More of his work can be found at ⊗ henry-wismayer.com.*

SUMMER OF SNAKES

Mhairi Quiroz-Aitken

Longlisted 2012 'A Close Encounter'

'This is bliss!' I thought as I stretched out in the hammock. The monsoon rains had been and gone for the day and the sun was now busily drying out the dirt tracks which had momentarily become rivers. The air was full of the gentle sounds of distant elephants and the steady background chorus of a thousand birds hidden deep within the jungle. Their beautiful songs were so persistent that they could be mistaken for silence.

I stared up through the canopy of palm trees to the perfect blue sky and watched as brightly coloured birds hopped from branch to branch. A tiny golden, striped squirrel jumped down on to the frame of the hammock and looked at me expectantly.

'I don't have any food for you!' I said.

The squirrel bolted along the top pole of the hammock's frame and leapt back up into the branches of a palm tree. 'Sorry!' I said.

But it was not me he was running from.

A black beady eye now peered down at me from the metal pole a metre above my face. The eye moved forward bringing with it a long, glistening, brown body. It slithered along the pole a few inches then stopped. Now there were two eyes and they looked down at me inquisitively. The snake's face was solemn and expressionless as it squared up to my own. Calmly I returned its stare. I smiled, remembering how differently I would have reacted just a couple of months earlier.

When I first encountered a snake I had been volunteering at the elephant sanctuary for just two weeks. Each morning I met with Chandra (one of the mahouts) and Rani (the elephant I had been allocated to). The days began with bathing Rani in the river – scrubbing her down with coconut husks, while having my own feet exfoliated by the swarms of tiny fish living in the murky waters. After Rani's bath was over we would begin collecting branches and leaves for her food.

On the day I first met a snake my mind had wandered far away from the tropical jungles of Sri Lanka. I carelessly pulled at branches and thrust my hand deep in among the piles of leaves. As I clasped hold of a few branches and began to yank them out I saw in among the rich and varied greens a flash of orange and red. Suddenly the bright colours shot up from within the branches and began dashing across the dry ground towards my feet. I let out a yelp and jumped back behind Chandra but instead of offering reassurance he sharply picked up the brightly coloured snake. With the animal now wriggling violently in his bare hands he held out the snake towards me. I moved further back. He came further forward. I ran. He chased me.

And then he began to laugh. I stopped, and feeling a little foolish, gingerly walked back towards him. He held the snake towards me once again and said: 'No danger, no bite.'

He put the snake down and I watched as it zigzagged across the ground and into the shade.

'That was not funny!' I said, but the mixture of adrenaline, relief and embarrassment made me laugh.

As the weeks and months passed these serpentine encounters became more and more common and just as the curry that had set my mouth on fire in the first week now formed part of my daily diet, so I came to regard these reptiles as entirely normal and harmless neighbours.

So as I lay in the hammock looking up at the shining black eyes and glistening brown body of the snake peering down at me I did not feel alarmed.

I heard Chandra approaching and prepared for another of his practical jokes. He never ceased trying to recreate the panic which he had found so funny on my first snake encounter, but I had become largely immune to his tactics.

He stopped.

I turned my head to see him standing perfectly still a few metres away. Silently he pointed at the snake and waved for me to come forward.

I laughed nonchalantly, and lay back on the hammock, stretching out my arms and faking an exaggerated yawn.

Chandra began walking slowly towards me.

'Come!' He said; 'Big danger!' He added pointing at the snake.

I looked up and was struck by the cautiousness of his movements and the wideness of his eyes. Either his acting was getting better or this was no joke. Suddenly I was reminded of the many miles between this hammock and my home and of the excitement but also the trepidation that I had felt during my first weeks in Sri Lanka. Once again the world around me became exciting and unknown.

***Mhairi Quiroz-Aitken** lives in Scotland. She loves solo travelling and her favourite destinations are China, Morocco and Colombia. These days she is mostly found closer to home with her two children, ponies, dog and tortoise. She can also sometimes be found in stand-up comedy clubs where she performs sets about artificial intelligence and data science.*

SNAKE TEMPLE

Sylvie Celyn-Thomas
Longlisted 2011 'Up the Creek'

The jungle is quiet. Empty, in fact. Just the vanishing shadow of a spider monkey and the tail of a monitor lizard sliding into a black pool, then nothing. The laughter of the Australian trekkers as they jumped up and down on the bridge has sent the animals skittering into the trees, taking any colour with them. Our footsteps follow a swollen creek and we peer into green darkness.

It wasn't like this fifty years ago, when my father was here.

He was a real jungle adventurer, a boy soldier, trekking through the undergrowth alongside Sarawak head-hunters. You had to keep your wits about you, there were terrorists then who could creep up and take you unawares as you took a morning shave. In his day, the jungle was a riot of noise, the air vibrating with the screams of monkeys and a medley of clicking insects and hooting birds. Quiet only came when the whole jungle stopped breathing to listen to the deep rumble of a distant tiger. On our early morning walk, all we hear are raindrops, big, heavy and relentless. Falling from those big black clouds of disappointment.

As the drops of rain sneak down my neck, I remember my father's tales of swimming in a clear mountain pool with his comrades, free for a while from army discipline, enjoying some rest and relaxation on the old colonial island of Penang. I imagine following the creek uphill to a waterfall and diving into the cool water to wash away the discomfort of incomplete wetness. But we just see green, nothing else. A dark, oppressive green.

I think about the temple that he had talked about. A red Chinese temple, almost swallowed up by the jungle, towering palms overreaching it. It was a long way from the grey chapels of Wales. A fog of burning incense and through the darkness, a pit of vipers, slumbering in the fumes. He spoke of Chinese worshippers standing entranced, placing their hands among the snakes in a hypnotic fervour. Inside my guidebook I have the tiny black and white photograph of my father beside the temple. Despite the strangeness of it all, he has backpacker nonchalance. Another day, another exotic destination. He's the sort of father anyone would want, with his batik shirts and slicked back hair, and tales of tigers and snake charmers in the Malaysian jungle. As we head back down the creek, I realise how much I want to find a trace of his past so I can bring a story back home to him.

The sky stays heavy as we travel down from the highlands, passing stilt houses and drenched paddy fields. The wetness and darkness intensify the green foliage and illuminate the colours of the durian, mangosteen and rambutan on the roadside stalls. The rain decelerates at last. The sky brightens, and puffs of evaporating clouds hang over the jungle.

We reach the busy airport road and are passing the industrial zone when we see the sign for 'The Temple of the Azure Clouds'.

The jungle has receded but the little temple is still there, vivid and gaudy now it has escaped from a monochrome photograph to reality. We walk under the dripping roof canopy, under carved dragons and enter.

The snakes too are still there, lazily draped around branches on the altar, intoxicated by incense. Emerald-green vipers with dangerous yellow stripes, they long ago slid out of the jungle to claim the temple built on their land and never left. The jungle may have quietened but the snakes still stake their claim to where it used to stand.

A photographer thrusts a handful of squirming snakes at us. I step back but he will not be happy until we have a family portrait with the temple's famous guardian spirits. I'm nervous of their flickering tongues and coiling bodies but try not to show it. The snake's muscles tense and grip my hand. Like a brave jungle adventurer, I smile for the camera.

This brave and poignant story was longlisted in the Bradt Travel-writing Competition in 2011. We have since lost contact with the author.

DOG DAYS

Peter Rimmer

Longlisted 2011 'Up the Creek'

The dogs were barking as daylight broke. It was always cold in the early morning and today was no exception. The sun took another hour to lift its head above the mountain tops. This was Mongolia in winter – sunny and dry, but cold, oh so cold!

Our dog teams sensed that they would soon be on the move. The sledges were prepared and positioned outside the gers at the bottom of the hill alongside the frozen River Terelj.

Four dogs were harnessed together and hitched to each sledge with the driver standing on the runners behind holding firmly to a wooden rail to steer. There were two breaking systems – the equivalent of an anchor, which gripped the ice to hold the dogs back, and a rubber friction pad. I never mastered the rubber pad but the anchor worked fine for me. The sledge was steered ski-fashion – knees bent, weight to one side or the other, and the dogs did the rest!

It sounds simple but on that first morning there was some apprehension and the adrenaline flowed freely as we prepared to get under way. We sped across the fresh snow, smooth as silk, the dog team pulling with all their might, tails up, ears back, tongues out, enjoying their run and chasing the team in front.

We followed the frozen creek avoiding cracks in the ice and the icy chunks pushed to the surface by the pressure as the river water had frozen rapidly in the sub-zero temperatures. The ice cracked behind us like gunshots, and the dogs slipped and slid on the icy surface where the cold winds had cleared the surface snow leaving a glass finish glistening in the bright sunlight.

Up the creek and along the valley, we sped high into the remote hinterland behind the town. It was exhilarating. Wrapped warm against the cold with four layers, two pairs of gloves, two hats and strong sunglasses; it left only the cheeks slightly exposed to the icy wind.

Away from the valley, we headed across uneven ground where we were on and off the sledges frequently. There were tracks everywhere – fox, wolf, deer, mountain hare and bear – but no signs of life beyond the tell-tale trails in the snow. From nowhere, a string of horses led by a young herder on horseback crossed our path and disappeared again into the forest as quickly as they had emerged like a ghostly spectre.

The frozen river was smooth for the most part but hazardous where the ice and rocks broke through the surface. These were the places where the dogs struggled most but they pressed on at pace heading towards the top of the valley, about 2,000 metres above sea level, with the sun dropping in the western sky. The overnight ger camp appeared in the distance and we followed a smaller creek, now frozen and snow-covered, winding through forest and woodland, across more rough terrain, and beyond the horses grazing before nightfall.

The camp had six gers and pens for camels, horses, goats and sheep. The head of the household greeted us – a Kazak man from the west with a wonderful set of whiskers and a glint in his eye. The young men were busy rounding up the horses that were proving troublesome in the presence of the dog teams; younger family members helped out with the other animals. We anchored the sledges, fed the dogs and settled into the family ger.

This was an extended family – grandparents, sons, daughters-in-law and grandchildren. More milk tea and silence. Mongolian families were reserved, at least initially. The children looked quizzically at their foreign guest. Our guide broke the ice in fluent Mongolian. A lively conversation followed and there was great interest in our travel adventure, the Ice Festival on Lake Khovsgol, and the Angil (Englishman) and his pictures from home. Our hosts appreciated a glimpse of our life – 'Where are you from?' 'What do you do?' 'How many horses do you have?' Small talk illuminated by photos of home.

The generator slowed as it ran out of fuel; a sign for bed. We remained in the ger with the eldest son, his wife and their youngest daughter plus eleven lambs and two baby goats! The fire was warm, the sleeping bags cosy and we settled down for a good night's sleep. There was not a murmur from the lambs or the baby goats in the ger. Nothing stirred, not even a sound through the clear, cold night air from the bears, wolves and foxes prowling in the hills. But we could feel their presence, and the paw prints in the snow next morning meant that we were not alone.

Peter Rimmer *is a freelance writer and photographer from West Lancashire. He was awarded a Master's Degree in Photography by the University of Bolton in 2013, and has self-published a photo book,* The Tide's the Very Devil, *about Morecambe Bay and its shrimp fishermen. Peter specialises in Para-Sports photojournalism and contributes to disability lifestyle magazines in the UK, USA, Germany and Algeria.*

A WILD NIGHT WITH BORNEO'S OLD MAN

Lewis Cooper

Longlisted 2011 'Up the Creek'

Our boat sloshes gently from side to side as our tour guide, Yaya, laughs. He corrects each roll with a flick of his head to check for traffic – there is none; this is the Kinabatangan River on the island of Borneo – and then a flick of his wrist to correct the growling outboard motor. The only traffic we are likely to encounter, he joyfully informs us, are crocodiles. As if to emphasise this casual observation, at the next turn in the tributary, one slips from a greasy bank and, is it just me, or is it submerged beneath the wake of our low sitting boat? We roll right and then left again. 'No worry Mr Andy!' beams Yaya and he steers us away from the river's rippling surface.

My girlfriend and I are deep in the rainforest on an expedition with a difference – to see wild orangutans. It is the brainchild of a local orangutan conservation charity in the area. Part of the cost of our stay goes to local communities like Yaya's who are working with the charity to help the orangutans that are left. They are so endangered that, although Borneo is one of only two places in the world where wild orangutans survive, there is no guarantee we will see them. Like seasoned zoologists, however, we travel in the hope that we might, rather than a certainty that we will, and anyway, a bed in the Bornean rainforest is alluring enough to convince us to leave Leeds and fly around the world.

What a welcoming bed it is too. The Kinabatangan Lodge is a stilted haven for wilting travellers. Comfortable and clean, the private rooms are complete with double beds, en-suites and ceiling fans, it's backpacking luxury after 27 hours' travel, but this is not the time to relax. The jungle comes to life as the sun starts to dip and we find ourselves travelling, two to a boat, chopping along the river with dedicated guides in search of the forest's famous old men.

Surrounded by nothing but the sound of the vast Kinabatangan floodplain there is an air of excitement as our four-boat flotilla moves slowly into a dark-mouthed adjunct armed only with powerful torches, cameras and passengers filled with butterflies. The humidity is of the constant, sweat-trickling type. Bushes, reeds and trees compete with each other in a constant chirruping of noise. High above us, monkeys squawk our arrival like a troop of waiting trumpeters and instantly branches bounce as birds take flight. The rainforest is watching *us* now. This is green tourism with a capital G and a capital T, this is how eco-tours should feel; fizzing, refreshing and stirring, just like the drink.

Our progress is wonderfully slow and hushed; outboards are fired up, then cut. We flop and float, snatching photos of proboscis monkeys and their hareems, zooming in and curling away from sleeping tree snakes and eyeballing the outline of yet more crocodiles. The guides talk hurriedly to one another and then to us, always informative, pointing, stooping and laughing at our 'ahhhs' and 'ooohs'. A sense of perspective descends as the rainforest flirts openly with us. Even without an appearance from its main attraction this is a sightseeing tour so rich it saturates the senses. A peculiar looking rhinoceros hornbill watches from on high, 'Za-zu!' shrills a beaming Yaya, sloshing our boat from side to side once more.

As dusk falls we retrace our route, full now with familiar faces, and our thoughts turn towards dinner's local delicacies – freshwater prawns, curried fish and a glass of something cold – when, in an instant, the mood changes. A guide is waving frantically ahead. The rainforest canopy is moving. Our collective heart rate rises. Our eyes widen to take in more light. Branches shake and snap. Something large is moving in the treetops. Yaya paddles us ahead of the pack and we wait. Shadows slowly take the shape of long limbs, branches blur into familiar forms and then out of the dusky light we see them, a mother and baby orangutan moving high above us, her blaze of orangey-brown unmistakable. We hold our breath for fear of scaring them and for a few enchanted minutes we inhabit their world, uninvited but in awe. We study each creaking move, savour every second in their company and quietly count our jackpot of blessings. As they fade back into the canopy that protects them, the significance of our sighting is measured instantly in Yaya's face – his uncharacteristic silence speaks volumes.

That evening our party's excited chatter adds to the constant clamour of the jungle. The lodge glows in the deep dark of the jungle night, our energies united by the old man of the forest, wherever he is.

This adventurous story was longlisted in the Bradt Travel-writing Competition in 2011. We have since lost contact with the author.

THE COLLECTION

Rachel Robbins

Highly Commended 2019 'Out of the Blue'

'Sunrise is best time to see him,' Wayan said. 'You want wake-up call?'

Eight hours later, a gentle tap-tap-tap on the cabin door wakes me from a deep night's sleep. Wayan is stood there with a banana in one hand and snorkelling gear in the other. He hands them to me with a toothless smile and points towards a dusty path behind the guesthouse.

'The north shore,' he says. 'Be careful of the collection.'

'The collection?' I repeat, wearily. He nods.

It is quiet, save for the caws of an enthusiastic cockerel and his harem of hens, which scratch vigorously at the ground. I walk past snoring wooden houses guarded by tethered cows whose eyes glint in the dawn's soft light.

The inland path comes to an end and I am on the eastern shore. Towering palm trees shelter beachside shacks that, from lunchtime, will sell bottles of Bintang to a soundtrack of reggae music.

The collection, I learn, takes place on a Thursday morning.

Hundreds of black sacks are piled on to the beach, the tide lapping gently against them. They bulge hazardously, as if teetering on the edge of a sneeze, threatening to burst and reveal their contents.

I stop for a moment, scanning the gargantuan amount of waste. The smell is pungent and my stomach lurches with a deep sense of helplessness.

I jog past the wall of polyethylene and continue towards the north of the island. The beach is uninhabited at this time of the day and its empty, seemingly clean sands are the antithesis of the eastern shore. I squeeze my toes together, feeling the soft coolness of the grains beneath my feet. I eat the banana that Wayan gave me, leaving the peel alongside my towel as I don my snorkel and fins. I run into the water.

A bed of dead coral dominates the shallows. Bleached branches lie lifelessly on the seabed and after a minute of swimming, I start to wonder if Wayan had meant for me to search here. But, through the dim orange haze, the white seabed starts to transform into a scatter of rocky outcrops and colourful sponges.

'Find the barrel,' Wayan had said the night before. Head down, I scour this underwater world for the tree-like structure.

I spot a solitary batfish up ahead. To my right, a butterfly fish swims by. A sea anemone dances in the tide, and I glimpse a flash of orange between its tentacles. I dive towards it and come face to face with a family of clown fish. A bold male approaches my snorkel with a furrowed brow, as if both curious about and irritated by the disturbance.

The sun's rays ricochet across the corals. From the cerulean of the ocean, the water's hues a vibrant Parisian blue macaron, I spot a flash of olive green.

Wayan was right.

There, cushioned on a giant barrel sponge, lies one of the island's oldest residents. An aged green turtle, its shell a hexagonal patchwork of earthy brown. I dive down, my ears popping as I sink deeper under the water. The turtle's round, black eyes roll lazily towards me.

'He's nearly as old as me,' I hear Wayan chuckling.

His golden head rests on the edge of the sponge while his fleshy underside bobs against the coral. A chunk of his left flipper is missing, a spotted edge fringed with a white scar.

I rise to catch my breath and the turtle follows. He swims smoothly to the surface and pokes his face out of the water to take an audible breath. With bubbles drifting from his nostrils, he returns to his sponge.

A hum of engines breaks the silence and I start to swim back to shore, conscious that boats cross these waters, quickly, from all directions. A stream of smoke rises near the eastern shore.

A plastic bag floats ahead of me, and I reach to grab it. A few metres beyond that, an empty can of Diet Coke – 'no calories, no sugar'. As the water gets shallower I realise that a sack has split and spilled into the ocean.

As I reach the beach I throw off my snorkelling gear and return to the water to drag the rubbish out. I forget that dead coral lines the seabed and I cut my toe. Blood trickles gently into the water.

I am startled by a cough behind me. A young local boy, cigarette in his hand, stares at me blankly.

I heave the waste on to the palm-lined path and pick up my banana peel. I go to put it in the bag, and then I hesitate. 'They take it over there,' the boy says, kicking the bag softly and pointing to the horizon. 'But it always comes back.'

***Rachel Robbins**' love of travel was sparked by a three-week trip to Ghana in 2012, where she relaxed on the Cape Coast and spotted elephants in Mole National Park. She subsequently lived in and travelled through Australia before eating her way across Asia and the Americas. Her fondest travel memories include swimming with marine iguanas in the Galápagos and hiking in the jungles of northern Myanmar.*

THE COLDEST MILES

Polly Evans

Winner 2006 'Taking the Road Less Travelled'

We had no idea how cold it was. Our thermometer recorded temperatures to minus thirty but the mercury had slunk well below its bottom-most marker. It was as though the liquid was, like us, hunkering down to conserve every last droplet of heat.

At night, we lay close together in our tent in an attempt to benefit from one another's warmth. Our sleeping bags contained not just bodies swathed in wool and fleece but the digital cameras, contact lenses and felt boot liners that we wanted to keep from freezing. On our heads we wore balaclavas and hats; each morning we woke to find an outer layer of white ice capped our skulls so that we lay like a line of stiffly swaddled pontiffs.

'Don't try to be comfortable,' said Stefan, our guide. 'It's impossible to be comfortable when it's this cold. The best you can hope for is a couple of hours' sleep.'

The dogs, like the mercury, were shrinking fast. There were seven of us and thirty-four of them; we were sledding along a part of the Yukon Quest trail that stretches for a thousand inhospitable miles between Fairbanks, Alaska and Whitehorse in Canada's Yukon Territory. We fed them snacks almost hourly – frozen hunks of fish, chunks of pork and chicken skins – but they were burning calories so fast they devoured them in a gulp, and still they grew skinnier. In the evenings we dressed the dogs in padded coats; during the day we strapped fabric booties to their feet so the ice wouldn't cut their paws.

We fitted the booties with bare hands – it's difficult to pull tight the Velcro that binds them when wearing gloves – and we ran to warm our fingers by the fire when the pain grew too searing. We were in no danger of being comfortable.

But then, each morning, after we'd been up for an hour or two melting snow for water, feeding the dogs, eating breakfast and dismantling our camp – the sun climbed above the pristine hilltops. The sky shifted in hue from pale cornflower to a deep, vibrant blue. And the temperatures, gloriously, rose.

Now we could look forward to eight hours of undiluted joy. On the fourth morning we ascended high into the hills following the tracks of a lone wolf. As we approached the summit, the trail became buried beneath fresh snow. My team struggled valiantly to bound through the powder but every now and then they lost their footing and plunged into the gulfs on either side of the packed path. For a moment they would disappear, submerged in the whiteness; then they'd leap to their feet, shake the ice from their faces, wag their tails and look round with a bemused expression that asked, What happened there?

By now I was growing to love these creatures and to know each of their characters. Sonar, in the lead, was diligent in the harness but at night she grew cold and miserable, and left her dinner untouched. Her partner, Pelly, was docile, biddable and happy to eat for two. Terror would chew off his booties at every opportunity but, when I came to replace them, he'd gaze at me so adoringly that all was forgiven. Val was light and lithe with a nasty habit of chewing the lines, while Log worked with an energy that was almost unfathomable. As we crossed frequent patches of glare ice he'd run so hard that he slipped but, even as he rolled on his back and turned to glance at me with horror, his eager legs would still be pumping. And finally there was Alex Large, loving and perpetually thirsty, he liked to snatch mouthfuls of snow as he ran.

We came down from the hills and continued through forests of spruce towards the Stewart River. Every now and then the dogs' heads would pivot and they'd accelerate wildly, following the scent of a rabbit or a grouse. It was dusk, the air was growing cold once more, and the time was ripe for shadowy thoughts. My mind shifted to the terrible tales I'd heard about moose: when they encounter a dog team they sometimes charge, causing injuries and even death.

'What should I do if I meet a moose?' I asked Stefan.

'Get a gun,' he said.

As darkness fell we arrived at the river. Small, snow-covered mounds peaked and dipped where, months ago, the freezing pockets of ice had jammed one against another. The moon was full and in its pale-grey light the dogs trotted in perfect rhythm. Just the patter of booties on snow and the gentle creaking of the sled disturbed that very silent world. It was deathly cold out there but I no longer minded for, in this remote wilderness, I had found a bitter heaven.

***Polly Evans** is an award-winning travel journalist and author. She is the author of the Bradt guide to the* Yukon *(winner of the British Guild of Travel Writers' award for Best Guidebook), and* Northern Lights: A practical travel guide. *She has also written five travel narrative books. She lives in Berkshire with her husband and three sons.*

5
THRILLS & CHILLS

"The butt of one of their rifles tapped politely on my hip bone."
Ella Pawlik

Despite our best plans, not all travel is smooth and trouble free. Animals, muggers, traffic, weather and landscapes are all determined to foil our plans. The best stories seem to emerge from the worst experiences. So, here is a selection of adrenaline-filled accounts: close shaves; dangerous moments; random humour and real peril.

Finland USA Colombia New Zealand Australia Vietnam Mexico Tanzania India Syria Kazakhstan Brazil Malaysia

IN DEEP SNOW

Cal Flyn

Winner 2013 'A Narrow Escape'

The waiting room of an accident and emergency department is as good a place as any to do a spot of sightseeing.

Here in Ivalo hospital, in the far north of Finland, the staff, patients and interior décor provide a great insight into the minutiae of Arctic life. Men in heavy furs nurse arms in improvised splints. Sami women dangle infants in tiny balaclavas over their knees, as snow melts and puddles around their feet.

On the wall a colourful display of fishing flies is revealed, on closer inspection, to be a collection of foreign objects removed from patients. 'Broken fishhook, pulled from eye,' translates Satu. 'Rusted nail found deep in flesh.'

It's a lively place to hang out, but you really need a good reason to get past the doorman.

Myself, I am bent and perhaps broken. I have been trampled by a horse; stiff in the neck, sore in the back and bruised in the buttock.

It's as good a reason as any. Strong blonde Finns – with shoulders wide as Atlas' – nod almost approvingly as I hobble through their midst, supported on both sides and dressed in dirty salopettes. I have joined the club.

Finally. I've been in Lapland for four months now, working for my keep somewhere they mush with huskies and sled with the native fjord horses. It's a tough life on the edge of the world, in conditions so cold that the tears will freeze to your cheeks as they fall.

I've been finding it hard to keep up with the sheer machismo of it all. Guns? I can't fire them. Axes? Can't chop a thing. I don't ski, I don't hunt, I don't fish.

But horses: horses I can do. I thought it would be my way in, my one useful skill. I'd train wilful Wilma, the stroppy fjord filly, to pull the sleigh, and that would be my 'in' into Lappish society.

And so it has proven, although not quite in the way I was hoping. When you train a horse, they call it 'breaking in'. In this case, Wilma was breaking me.

It happened so quickly, I didn't feel pain until I found myself lying limp in the deep snow. One second we were calmly backing her between the traces, buckling the harness, stepping her forwards; the next was a tangle of limbs and yelling and the rushing and crushing of hooves.

Wilma panicked, I was later told, during the long and painful drive to the nearest hospital.

She bolted forwards, trying to shake off the sled, but as it was tied on tightly it kept up a close pursuit. I pulled at her bridle but only managed to swing her round to face me, before she knocked me to the ground and galloped over my body, pulling the sleigh across me for good measure.

Fjords may be short in stature, but they are strong little horses. And heavy – half a tonne, at least.

When Wilma had finally torn off the sleigh and stopped – just short of the frozen lake, thank God, thank God, thank God – my friends ran back to check on me.

'I'm fine,' I said, trying to sound casual. I was not fine, clearly. But I could roll over, very slowly. And, with a great deal of help, I could stand. I'd left a cartoonish imprint of my body in the drift, arms flung high – a snow angel in distress. But where my waist should be, deep tracks from the hooves and the sled.

'You were lucky to fall in the deep snow,' said Erki, my Finnish boss, as I limped away. 'The ice on the drive, here. It is hard as concrete. It is only inches away. Your insides, they would be very crushed.'

After a three-hour wait at the hospital, the doctor agrees. There is blood in my urine, he says: my kidneys are bruised. But it could be worse. On the balance of horror, as Erki would put it, I have come out very well.

I return to the waiting room to await my prescription. Two men are pulling on heavy layers, ski masks and camouflage-print snowmobile jackets. They nod at me, almost in recognition.

Cal Flyn *is an author and journalist from the Highlands of Scotland. She has worked for the* Sunday Times *and the* Daily Telegraph, *and contributes regularly to* Granta, The Guardian *and* Prospect. *Her 2016 debut,* Thicker Than Water, *was a* Times *book of the year. Her acclaimed second book,* Islands of Abandonment, *about the ecology of abandoned places, was released in 2021. She was made a MacDowell Fellow in 2019.*

SUMMER IN THE VALLEY

Sylvia Dubery

Finalist 2011 'Up the Creek'

The desert highway quivers in the heat haze. I keep one eye on the petrol gauge which quivers above zero, as if watching it will stop it from dropping to empty.

Summer in Death Valley. Inside the car, we are cocooned in a cool air-conditioned microclimate which is draining the petrol away but outside is a furnace. Every road sign is a metaphor for the hopelessness this land offered the ghosts of doomed pioneers. Funeral Mountains, Dead Man's Pass and Last Chance Canyon. It's like a devil's word association game with an edge of reality because if we run out of petrol, we're really up the creek.

Furnace Creek at dawn. All night long, a fierce hot wind has blasted our balcony but the morning is still. Before sunrise, we swim in the pool, still bathtub-hot from the previous day. The sun races up from behind the ochre hills and it's as if someone has flicked a switch to instant heat.

We buy coffee and cookies for breakfast and drive to the Devil's Golf Course. Sipping steaming coffee, we tread among the miniature white pinnacles, listening to the salt crystals crackle like a breakfast cereal. I crouch down to photograph the evaporated lake floor and see little snow-capped mountains through my lens. We stand on an ocean bed at Badwater, 282 feet below sea level. Tiny fish defy gravity in the salt-saturated shallows, bobbing around, going nowhere in their last dregs of sulphurous ocean. The heat slows you down and we simply forget about filling up the car.

We explore a trail through a valley landscaped by heat. The extreme climate has eroded the rock into smooth contours in an artist's palette of colours, layers of burnt umber, oxide red and raw sienna lined with pale scrub flowerbeds. It has sucked moisture out of the earth to produce concrete-hard mud flats. We stand on ground that is dried out and cracked, a mummy's skin, drained of its colour and vitality. Mysterious moving stones shuffle around, leaving a mud trail behind them, though they only move when nobody's looking. A bit like that petrol gauge.

At Stovepipe Wells, there is a last chance to refuel but we are distracted feeding the birds from our water bottles. Little brown starlings and a desert red oriole peck gratefully at the splashes of moisture that we scatter on the ground. We climb the scorching dunes and run back down, our shoes heavy with sand. We photograph sandblasted driftwood and follow the tracks of a fleeting coyote. And all the time, the temperature is rising.

Common sense just slips away, until we are on the long desolate stretch of highway, heading west, speculating on how many miles are left once the fuel light glows. There is nothing but dry rubble and bleakness for miles. Ahead of us, the Panamint Mountains form a formidable wall, just as they did for the Forty-Niners, the pioneers who crossed the valley on the gold trail. Our transport is more reliable and hardy than their mules, so long as we keep it well fed.

At last, we begin to rise from the valley floor and some Joshua trees appear, their spiky stunted arms reaching skyward. I want to photograph my husband standing beside them, frowning into the sun like Bono.

'No,' he says, eyes ahead, 'Too risky. It'll use too much petrol.'

I watch the trees pass with regretful longing and store the image in my memory instead of camera.

At last, far in the distance, we see Panamint Springs, a lonely mirage in the barren landscape. We drive in silence towards the yellow signs, only breathing again as we reach the lonely, unmanned gas station. We have made it; together we have triumphed over the elements. Other cars are stopping, more relieved faces, laughing now that the worry is over.

We snatch up the nozzle and slip in a credit card.

The card is rejected.

There is nothing to be said. We turn off the air conditioning and keep driving.

Sylvia Dubery *comes from the island of Anglesey in North Wales and grew up beside the sea and mountains. She works as a teacher in a first school but likes to spend summer holidays on archaeology digs and travelling. She loves visiting ancient ruins and her favourite place, which she keeps going back to, is Pompeii. She loves the drama and adventure of deserts too, but would definitely recommend being better prepared for Death Valley.*

A PESTLE AND MORTAR IN PARADISE

Ella Pawlik

Finalist 2012 'A Close Encounter'

I returned from the Lost City. I nearly didn't.

It was the rainy season, and the torrents of water that poured down the mountainside were washing the road away. Our jeep ascent into the Sierra Nevada involved driving with one wheel overlapping the edge of a sheer precipice. We had two contra guerilla guards accompanying us in the truck and the butt of one of their rifles tapped politely on my hip bone. Climbing out of the jeep at the top, we met a group that had just finished the trek. They looked haggard and most definitely worse for 'where'. We asked them how it had gone. One man thought long and hard, and replied:

'It's like meeting with the devil.'

And so we set off. Uphill and sweaty – we were floundering mounds of uselessness. Our guides didn't seem to notice. We finally made it to the top and spent the first night sleeping in hammocks under the stars.

We rose with the sun and saw paradise. The butterflies were incandescent and bigger than birds. The bromeliad-lined trees were colossal, and there were flowers so big they looked like fruit. I fell in love with world all over again. And again. And again. And I didn't stop until we stopped walking. On the second night we camped right next to the ferocious Buritaca River. Slowly but surely we were working our way into to the obscurity of jungle. Mosquitoes and sandflies became squadrons of bloodsucker fuckers.

The next day we began the final stretch to the Lost City. On the way, we swam in a crystal pool with waterfalls, and I plotted a graph of no over yes. Then we had to cross the big river, the ferocious one. Seven times. The current was so strong we used ropes to get from one side to the other, and we blobbed around like cumbersome blubber balls while our guides balanced expertly on nothing. Safely on the right side, we soggily approached the stone steps to the Lost City, and climbed all 2,000 of them.

There. It was beautiful. Crumbling stone circle terraces rose implausibly out of the mountains like rejected offerings to the clouds. Serene. Detached. A whole land, a whole civilisation built on the dreams of dawn. And the jungle had reclaimed everything. We spent the afternoon in dappled sunlight, exploring what already was. Rising with the sun again on the fourth day, we toured the ancient city with our guide. Three snakes and a toucan later, we began the long trek home.

It turns out the devil was in a tree root. I tripped over it, and fell on to rocks below. Lying with my face in the mud, and my knee and the rock like a pestle and mortar, all I could think of was the word 'splat'. I briefly saw my pristine white kneecap, and then the blood started. With no option but to continue, I dragged my leg behind me, also trying to ignore the stabbing pain in my side. At base our guide boiled some water and salt to try and clean the hole in my knee, although he forgot to cool it first. I couldn't cover my leg up so the mosquitoes and sandflies feasted on it. At one point, a moth got stuck in the wound.

I had to conquer some deeply unforgiving inclines the next day. By the time we reached base for our last night my leg was so swollen it was starting to blend in the with ancient jungle trees. Waiting for the jeep to come and collect us, I noticed the swelling had spread up one leg and was creeping down the other. Both my legs were shiny and purple and it was becoming hard to tell where the joints were. The precarious drive down didn't seem nearly as scary this time as I was concentrating on trying not to let my legs explode. We got back to Santa Marta in floods – there was no way of telling where the sea ended and the road began. A pathetic fallacy of grey.

In hospital, I was treated for a cracked rib and cellulitis. 'Just in time.' Apparently. I had an antibiotic shot from a needle so wide it felt like a gutter, and was given millions of pills to lay my head on. For a week after, I dragged around swollen, fluid-filled legs – thinking of the graph I had plotted and waiting patiently for the infection to subside.

It was worth it. I saw what should and shouldn't have been lost. I would do it all again. And, it would appear that I took the Satan bullet for the whole trekking group. Everyone else was completely fine.

Ella Pawlik *writes words and people pay her for them. She's a Bristol-based travel writer, journalist and copywriter. Her love for exploring this beautiful planet has led to a life spent dedicated to preserving it. Keen to make sure that future generations will also have the opportunity to fall over in verdant lost jungles, she writes to inspire positive social change and encourage people to fight the climate emergency.*

THE DISAPPEARING BEACH

Simon Duncan
Finalist 2011 'Up the Creek'

'Oops,' he says.

I laugh at my brother. We are in his car which is now perched on top of a mound of black sand.

'OK, try again,' he says, slamming the car into reverse and throwing us back down towards the tarmac road. He revs the engine and points us squarely at the goal. The car roars as we leap over the hill and on to the empty beach on the other side.

'There!' he announces triumphantly.

You can drive on certain beaches in New Zealand. We're taking one of these 'beach highways' on a stretch of the west coast from Muriwai to Kaipara Harbour, a vast natural and empty harbour surrounded by forest, sand dunes and a few 'batches' – summer homes – peering down from the hills.

'The same driving rules apply on here as on the roads,' he tells me as we set off.

'Even speed limits?'

'Yeah, I think it's sixty.'

I glance over at the speedometer. We're doing double that.

Fortunately, the wide beach is deserted in early autumn. To our left the Tasman Sea laps calmly at the shore. The trick is to drive near the water's edge where the sand is harder and the tyres have more grip. We tease the waves with our wheels. On our right a desert of sand dunes loom and behind those, the dark dense forest hides all manner of imaginings.

As we head north the volcanic sand fades into the familiar gold of postcard pictures. It is quiet, just the slosh of tyres and the smell of the sea. We have the freedom of the open beach. On inland roads this journey would take nearly three hours; the beach highway reduces it to forty minutes.

We reach the harbour and leave the car. From a distance it looks even more incongruous; a lonely metal blob marring the landscape. The harbour is renowned for shipwrecks. Over the centuries sandbars have lurked beneath the surface to topple unsuspecting explorers. Now they are exposed and we hop from one to another. We are kings of soft small islands. Pools have been created in the low tide and we nudge the trapped fish with our toes and irritate the flitting colours.

My brother and I are intrepid explorers again, an echo of our childhood holidays. We reminisce and exchange news from our lands half a world apart; we poke and prod around the harbour; renewing our bond. He is tanned from a summer spent outdoors. We compare arms and he sniggers at my bleached skin. His hair has been made golden by the sun; mine is as dark as an English winter.

Clouds begin to brew and move in from the west. It is not long before the first fat drops of rain fall. We retreat to the car to feast on the stash of food and drink. The rain beats out its tune on the roof.

'The tide is turning,' my brother announces through his sandwich.

'Really? Already?' I peer out on the rain and grey.

'The sandbars have gone.' He points and I follow his finger. There is now only water in front and, I notice, around us. Here at high tide, he tells me, the beach submerges completely.

We make our escape. As we drive I notice the beach is narrower and the waves begin to nibble at our wheels. Shallow trenches have formed and we slow down to ride over the softening sand.

'So what happens if the water cuts us off?' I ask.

'We'll have to drive as far as we can up the dune and sit it out.'

'For how long?'

'About twelve hours,' he says nonchalantly. 'But we'll be fine.' He jerks the wheel and we skid away from another encroaching wave. I'm unconvinced; there is no beer in the car for a start. It would be a long night. We cannot get trapped. The race is on.

We try to keep a constant speed. We cannot slow too much in case the car begins to sink. The tide presses in, pushing us towards soft sand on which the tyres cannot grip. More water slaps the wheels but the dunes recede, giving more space, and we outwit the water with speed.

We reach the gap in the dunes and spin around in the surrounding pools to rid the sand from the wheels. My brother takes aim at the hill.

'I want to do it!' The petulant child rises in me. We swap seats and I prepare a long run up.

'Right then, you need to…'

But I already have the pedal to the floor, flicking wet sand everywhere. I don't let up though the hill looks like a solid wall. We whoop like victorious cowboys, flying over the sand, away from the disappearing beach.

Simon Duncan *made a commitment to himself to spend every birthday somewhere different. This has had the unintended consequences of enduring violent tropical storms on Easter Island, bobbing along the Dead Sea in 107-degree heat, and gasping for breath in Cusco because he thought altitude sickness was for wimps. He still stands by his decision.*

FIVE GET CLOSE TO HEAVEN (WITH APOLOGIES TO ENID BLYTON)

Peter Hurley

Longlisted 2013 'A Narrow Escape'

On that fateful New York City morning in September 2001, I stood in the bright sunshine on the open observation deck atop the World Trade Center's South Tower.

With me, were my wife and three young sons: 12-year-old twins and a 9-year-old. We were at the start of a cross-continent trip. After two days in New York we planned to drive over 3,000 miles west to Los Angeles.

With so much to experience, during our brief New York stay, we had set out early and excited that day.

Breakfast was supplied by a mid-town Korean fruit vendor on Broadway. It was consumed on the packed and grimy, diesel-belching bus winding its halting way south towards Manhattan's financial district.

At 8am I traded what seemed an exorbitant amount of dollars for five tickets. A minute earlier, passengers and crew aboard American Airlines Flight 11 had buckled themselves in and departed Boston's Logan Airport, for what they assumed would be a 3½-hour trip to Los Angeles.

The cover of the official World Trade Center leaflet opined that being on the roof of the building would be: 'The closest some of us will ever get to heaven'. The tragic events that followed probably proved prophetic.

The five of us, along with a few other eager tourists, left our breakfasts behind in our stomachs as we were whooshed upwards in the express elevator, at 27 feet a second. We all stared up silently, seemingly hypnotised, as the ascending floor numbers blinked on and off faster than we could count them mentally.

Eventually, we all stepped out on to the open rooftop's observation deck. Despite the early hour the air was reasonably warm in the late-summer sunshine. Above us, a virtually clear bright blue sky was dotted with a few white clouds, like floating mounds of fluffy mashed potato. A reverential silence descended upon us all, as if the location was a place of pilgrimage. It was soon to become one. The prevailing stillness was in complete contrast to the cacophonous sound of street-level Manhattan we had escaped, bustling a third of a mile below.

Our leaflet advised that our 'small' admission fee would provide – on a clear day – a view of five of the fifty United States: New York, New Jersey, Pennsylvania, Connecticut and Massachusetts.

We initially gazed north from our dizzily heighted vantage point. Doing so we could clearly see the Empire State Building jutting above the other soaring mid-town Manhattan buildings, three miles away. Beyond that, it all looked fairly flat. We certainly couldn't distinguish upstate New York from Connecticut or Massachusetts. Had we been able to do so, we might have seen United Airlines Flight 175, with its crew and passengers, commencing its airborne journey from Boston. Flight 175 was to have the same ultimate destination as Flight 11.

We intended to catch the ferry to Staten Island at 9am. However, at the rate we were going, I wasn't sure we would make it. There were another three sides of the building from which to view.

We looked east over the imaginatively named East River, and across Brooklyn Bridge towards, well, Brooklyn. 'That's where two of my great-aunts came from London to live with their husbands,' I enthusiastically advised my non-enthusiastic sons, who showed no display of curiosity.

We sauntered across to the southerly view. I identified Ellis Island and the Statue of Liberty to my sons. 'That's what millions of immigrants to America would have experienced when they first arrived,' adding redundantly: 'including your great-aunts.' Interest in their great-grandmother's sisters remained unpiqued.

Time was moving on apace. It was 8.30am. If I had known about them beforehand, I might have wondered where Flights 11 and 175 were at that very moment.

We moved around to check out the westerly – and last – view. There, across the Hudson River, was New Jersey. At last we were able to distinguish another of the five promised viewable states. Pointing across the immediately flat, western, landscape, I said: 'We're about to explore 3,000 miles of America in *that* direction,' adding firmly, 'And we'd better begin now!'

Five minutes later, at 8.45am, we stepped satisfied from the lobby of the World Trade Center South Tower. We were, of course, completely unaware that – at that very moment – Flights 11 and 175 were being piloted towards their targeted objective.

Within sixty seconds our ears were suddenly pierced with the terrifying sound of crashing, crunching, mangling, metal – quickly followed by screams and shouts. We looked and saw that a white delivery van had hit a yellow cab.

For the five of us, that fateful day in New York City was on Monday *10th* September 2001.

Exactly 24 hours later, the world changed.

***Peter Hurley** was born in London in 1948, but has lived in Wiltshire for the last thirty years. He has an MA in American Political History. He has retired from running a training and business consultancy. He has travelled extensively in the United States and Europe. He has been compared to Bill Bryson – because of their similar physique. The wife and three children mentioned in his story retain their relationship with him.*

WOMEN DRIVERS

Sally Watts

Finalist 2005 'If Only I'd Known'

The signs were not good. Steam billowed from the engine and an ominous hissing noise with occasional 'clunk' was clearly audible under the bonnet. We were 100 miles on to the Nullarbor Plain, two females, mechanically clueless, circumnavigating Australia in a van. With true female logic we cranked up the radio to hide the noise, accelerated and headed into the unknown.

We had named the van Lionel, after Lionel Ritchie. We would sing along to his catchy tunes as we drove across this vast country sometimes wondering if we would ever see a human being again. Sadly the tape self-destructed halfway across the Northern Territories but by then we knew all the words.

Fate was kind to us that day. We were 20 miles from a roadhouse; an oasis. In the weeks to come, these garage/shop/bars became our link to the outside world, a telephone, a bathroom, cold beer and chocolate. I think we stopped at every one. We would park Lionel in the shadow of a road train. These huge multi-wheeled vehicles appeared like monsters on the horizon. Sometimes we spotted them half an hour before they reached us, the driver waving and the horn blaring and even after several weeks we would hold our breath as they passed us leaving Lionel shaking in their wake.

Now, as we drew on to the forecourt the sight and sound of our ailing vehicle had attracted a cluster of people. We emerged from the van hot, sweaty and very relieved.

A large man dressed in filthy jeans, checked shirt and baseball cap sauntered towards us.

'G'day,' he said. 'You won't be going far in that today.'

He pulled a thick rag from out of his trouser belt and opened the bonnet, releasing a cloud of steam and the stench of burning oil.

'It's your big end,' declared Wayne several minutes later.

It meant nothing. We looked suitably glum guessing from his demeanour that this was not a good sign.

'Can we get the part?' I asked optimistically.

If only I had known some basic car mechanics I wouldn't have felt like a useless female. His face revealed that this was exactly what he was thinking. I tried to disguise my English accent – being a Pom would have been the icing on the cake.

'You need a new engine – might get away with a recon.'

I wasn't going to ask what that was.

'Sorry girls, could take a couple of weeks to sort this out.'

Wayne looked triumphant; we looked miserable. Dreams of kangaroos, sandy beaches, Ayers Rock and wild adventures flashed before me.

'No worries,' he grinned, 'there's a room out back you can have.'

I look at Aly, my fellow traveller; words were not necessary. We could not spend two weeks in the middle of nowhere with Wayne. We raided our emergency funds and headed off to the bar for a beer.

We had spent the last few months in Sydney working and saving until we had enough money to buy Lionel. The 'extra' mechanical check-up was too expensive so when a friend glanced under the bonnet and said 'looks fine', that was good enough for us. We threw our worldly possessions in the back and, clutching a list of names and addresses of friends of friends, we left the city. Our combined auto knowledge extended to filing up with petrol, oil and water. Our navigational skills were no better but as the map showed a road around Australia, we were happy. We were heading for Perth, it was 1984, the Americas Cup had put it firmly on the map. It was the place to go.

'Sounds like you girls need some help.' We looked up into the eyes of our saviours, John and Dirk, two scruffily dressed locals. They had to get to Perth but had no money: perfect. Over a couple of beers we struck a deal. We would pay for petrol, food and beer and they would tow us 800 miles across the Nullarbor Plain. Aly and I looked at each other and without a flicker of hesitation we agreed. We were back on the road.

Next morning we hitched a beer-laden Lionel up to an equally unreliable-looking station wagon and headed to Perth.

We made it. Over the next few months we had a recon engine and a fan belt fitted, eleven flat tyres, a faulty petrol gauge and two broken headlights repaired and finally locked ourselves out of the van in Brisbane.

Back in Sydney we bade a tearful farewell to Lionel six months and 20,000 miles later. I've often wondered what a different experience it might have been if we had known something about cars – but no regrets.

Sally Watts *is a practice nurse and aspiring writer, and has recently completed her first novel. Travel is a common theme of her writing and over the years she has combined nursing with travel – working in Australia, India, Madagascar, Guyana and Canada. She lives in London with her husband, dog and campervan – the latter, always packed and ready for an adventure.*

EYES CLOSED, FULL SPEED AHEAD

Claire Morsman
Finalist 2012 'A Close Encounter'

The searing temperatures of the mid-Australian outback sap my energy. Distance has taken on a whole different meaning and although the Stuart Highway stretches ad infinitum, seemingly an unchallenging drive, the monotony of the featureless landscape and intense heat throws down its gauntlet. The endless pinky tarmac hums beneath my tyres, soporifically willing me to let my eyes close.

Just for a moment. Close.

Alone in a slow-cooking van of the very highest backpacker quality, I put another tape into the slot, all windows wound down flush to the bodywork, mosquito net flapping like a sail in the back. The vermilion dust all around – on everything I touch, in everything I eat and in every crevice of me and my van.

Sleepy. Close. Just for a second.

I turn the music up higher and ya-hooo out of the window. I pull my head in quickly though, as, in reality, the experience is like staring into a hairdryer, with the setting stuck on Too Hot.

Sleepy. Hot, heavy eyelid rims.

The horizon is always a simmering haze and the road ahead is ruler drawn. I fail to notice the devil's own transport behind me until the roar fills my own vehicle and his chrome bull bars loom through my back window. Fifty metres of road train pass in a shattering rain of dust and pebbledash. The top of his wheels align themselves to my roofrack. Shiiiiit!

Go on. Close. One second won't matter. Eyes will feel less itchy and heavy if you do.

I pour some warm water out of a plastic bottle on to my head, rub my eyes, and then see somebody on the barren dust. A hitchhiker. I find it strange to have to drive the van, actually slow down, change gear and stop.

He is scruffy, Irish and bothered by the flies. I pull my shirt away from my sticky back. Off again. Soon, I notice his eyes closing and catch him looking longingly at the mattress in the back. 'I'm knackered,' he says. 'I'd be more help if I get some kip now and drive later on. Refreshed like.' 'Fine,' I nod, inwardly un-fine. *Closey closey eyes. Go on…*

A terrible stench fills the van. The hellish aroma wakes me up. We whizz past a swollen bull, inflated like a Goodyear blimp and heinously rotting in the heat. Nightmarish birds hover for a feed. Entirely awake now, my right arm burnt to a crisp to match the outback hues, I notice black clouds on the horizon. Rain? Surely not. The cloud is moving fast and I anticipate the novel movement of flicking the windscreen wiper switch – until I smell the smoke.

As acrid as the decomposing innards of the bull that had come off second best to a road train; yet this smell is different. It's the wafting stink of impending death to all those who cannot run fast enough. I consult my rear view mirror, hoping that this might be the moment when my co-pilot might snap into action. What happens in a bush-fire exactly? Everything burns! Calm. It's a long way away yet.

Unlike any other outback vista, a wall of fire moves bloody quickly. I call out. 'Ahem, hello! Jeff, er Julian.' What is his name? 'Christ!' The flame is gorging on the scrub to the left of us. Surely we'll be fine? It's all around! It's on the road! The ROAD is burning! Smoke is stinging and blinding my eyes, by now ineffectually wide with terror. Sleep, except for the eternal kind, is now far from my mind.

'The tyres are melting!' I scream in absolute sheer panic. I feel the van give in to the sluggish pull of the bubbling tarmac. My feet are bare. I'll burn to death if I leave the van; I'll burn to death if I stay put. We're slowing. My hot tears of panic help clear smoke from my eyes and then I feel a hand on my shoulder. 'Let her out!' his accent commands. I push my foot down and we surge forward, hurtling through the wall of all-consuming flames. His voice murmurs: 'Now we're sucking diesel,' as he turns back round to find a comfortable spot to sleep again.

Rigid with shock, I drive until we reach a feral outpost. The wheels hadn't been melting. It'd been the crackling inferno surrounding us that had sounded like burning rubber. We stop. My unflustered, well-rested companion climbs out. He mentions that the fire had been good craic. Would I like him to drive for a bit?

I move into the back and find that I can't sleep. Eventually I push the visions of death by incineration to the back of my mind and persuade my eyes to *close*, the endless pinky tarmac humming beneath the tyres.

Claire Morsman *used to travel extensively, although this has recently been curbed by lockdowns and small children. She is an international English examiner, professional declutterer, Moroccan guesthouse owner and Morsbag maker. She is always looking forward to visiting somewhere new, and often longing to revisit somewhere she's already been. As long as she's on the move, she's happy.*

THE CHASE

Lauren Hatch

Unpublished Writer Award 2013 'A Narrow Escape'

It's 4am and I'm running for my life.

I'm running faster than I've ever run before. I've forgotten that I hate running, can't do it – don't do it – none of that matters right now, only closing the distance between me and him: he is escaping me.

A continuous drone of neon lights provides an atmospheric soundtrack to my chase. This is accompanied by the flip-flap of bare feet pounding pavement and the boom-skip-thud of my heart beating in my throat. I appear to have lost my shoes. I appear to have lost many things – my mind being the first and most obvious victim.

Only a shopkeeper up a ladder witnesses my chase. 'Why is he up a ladder at this time in the morning?' thinks she who is chasing a young man down the street at 4am, shoeless. He turns a blind eye. He's more concerned with the string of bright paper lanterns he's fixing above his shop front. They trickle out of his hands like candy-coloured pearls spilling from a jewellery box.

I don't labour on this, I can't. I haven't got a moment to spare. You see I'm in pursuit of him, or more accurately the bag that is in his possession: my bag. You see my whole life is in there – my passport, my money, my diary, my phone – my everything. And he's getting away with it. Fast.

Yesterday I boarded a bus in Hue and embarked on an 18-hour journey squished between a chicken coop and an overstuffed backpack, before being ejected right here in the street, apparently arriving at my destination. This seems to be standard in Vietnam right now. It's only been a few years since it flung its doors wide open to tourism and there isn't much choice, you just accept what you're given.

Only five minutes ago I had perched on the edge of my backpack leafing through the pages of my guide and feeling hopelessly lost when, a young stranger approached me asking for a light: 'A light? Do I?' I'm not sure but I rummage inside my handbag on the off-chance. And in that second, that ever so obliging moment, my bag was gone. A whoosh, a gasp, a flip-flop hits the pavement and the stranger has run into the night with my handbag and my life.

So here I am hot on his heels like a sandfly at dusk, when suddenly mid-chase, something entirely unexpected happens.

He stops. He waits. He turns to look straight at me – the thief meets his prey. Silence fills the air like a raincloud fit to burst and a question mark hangs over which one of us will make the next move, who will be the first to break? Yet I know, at least I think I can sense, that he means me no harm. He is ever so, ever so young and his expression now appears less belligerent youth and more bewildered child. It tells me he didn't expect it to get this far and he has no idea what on earth he'll do next. So, he does the only thing he can do – he flings the bag back to me and then whippet-quick he disappears into the night.

And that is that. Within seconds the silence has broken, the sun has come up and the unknown street is beginning to stir. A throng of motorbike engines purr off into the distance, while more shop owners appear, with more ladders, more lanterns, more pearls… the day begins.

Later that morning I sit above one of the shops in my room watching the pearls and the chaos unravel beneath me. Kim Lee brings slices of dragon fruit like some offering from another planet: all pink, prickly and other-worldly. 'I got my life back,' I rejoice. My host narrows her eyes like an overprotective parent, 'you could have lost your life.'

I nod in agreement, but I barely believe it. Instead, I think that the boy is just like the city: a little hard around the edges and slightly wary of the tourists invading his world, but there is kindness there – I saw it – a readiness to stop for a complete stranger and let them in. And over the next few weeks I soften too. I let the city in, right under my skin, and then I grow to love it.

Nowadays I run for the hell of it – 5k, 10k, a half marathon... But I've never run as fast as I did that morning in Ho Chi Minh. I doubt I ever will.

***Lauren Hatch** is a writer who lives in Hove. She first caught the travel bug when she visited New Zealand with her father as a teenager, a trip that inspired many more adventures, including a year spent backpacking around the world. Lauren also writes for charitable causes and loves paddle-boarding and going on bike rides with her family.*

ZAMA

Jack Losh

Finalist 2013 'A Narrow Escape'

The dozen bodies around me are motionless as I wake. Dawn still feels distant, the air cooler now, the courtyard outside dead. I pull on a T-shirt and stash a few dollars in my pocket for potential bribes should I get caught. Someone lets out a grunt as I creak open the door and descend the wooden stairs into the pitch-black Mexican night. Picking up a bike, I head southwards to the cliffs.

I had arrived the previous afternoon after four days of living a double life in Playa del Carmen, further up-coast on the Yucatán Peninsula. By day, I felt like just another gringo on the beach, never more than a few steps from a Starbucks, Burger King or Gucci outlet. But by night, I was among the locals – the hotel cleaners, taxi drivers, café waiters – sleeping in the suburbs at a friend's modest whitewashed house. Time had come to take to the road again. Fifty miles later by cramped *collectivo* minibus, I alighted at Tulum, a dustbowl of a town, but home to some of the Mayan civilisation's most spectacular ruins, set high above the Caribbean Sea.

The walled, time-worn city – populated by over 1,500 inhabitants at its peak half a millennium ago – stands atop 40-foot cliffs, gilded by virgin sands and white surf below. El Castillo and the God of Winds Temple dominate the view, marking the dramatic seaward entrance to one of the last cities built by the Mayans. Before invasion and Old World disease, this fortified town and trade hub flourished, exporting obsidian rock, jade and artefacts of gold, with salt and textiles arriving by sea through a gap in the reef. It was originally called Zama, the 'City of Dawn'. And it was this name, this promise, that had brought me here as the town nearby still slept.

Half an hour before sunrise, I make it to the beach after pedalling down a rough dirt track, tunnelled through undergrowth. It would be hours till the site formally opened but a brief recce the previous day had revealed another way in. Cliffs beneath the ancient city rise at the northern limit of the bay, past wooden huts and small fishing boats moored offshore. I reach the foot of the dark wall of rock and, with only the light of my phone to guide me, begin edging along a thin ledge above the inky Caribbean waters, invisible waves breaking close to my feet then sucking back hungrily. After slowly traversing a few hundred metres, I make it to a slanting bluff and clamber up to the foot of El Castillo, the clifftop fortress that guards this dead city.

First light filters through the charcoal clouds, casting the temples and turrets in a ghostly grey. No soul is here, at least none belonging to the living. The manicured lawns are empty, the stony footpaths silent, as I try to glimpse Tulum's half-imagined denizens through squinted eyes. On the far north of the site, I spot the God of Winds Temple, the highest lookout point. Criss-crossing past empty ruins and solitary palms, I reach the isolated building and take my place, back to the seafront wall. Cold, blue light spills over the Caribbean and on to Tulum, illuminating this fabled City of Dawn. I alone may be enjoying the sight this morning but I remind myself that many eyes before mine – of Mayan sentries, traders and conquistadores alike – have taken in this same, serene sunrise.

Within half an hour, a full sun is above the horizon, though obscured by an increasingly unsettled sky. I head back slowly towards the cliffs to make my stealthy escape, taking in the grand emptiness of the ghost town along the way. Then a yell to my right. Just a hundred yards away, two guards spot me. I break into a sprint and they take up the chase. Reaching a gap in the fence, I quickly crawl through, losing my flip-flop which I snatch off the dirt. Behind me the guards are fast gaining ground, El Castillo's imposing façade rising high above them. Turning, I rush towards the cliff face, scratching my legs on thorny scrub, and scramble down. One guard keeps up the pursuit, falters, then gives up.

Heart pounding, I eventually reach the bottom and place both feet on the beach. A few solitary figures have surfaced now – fishermen, early-morning joggers, café owners. The sands stretch out beneath a tenebrous sky as a day of drizzle closes in and I make my way back to the roadside hostel, replete with the memory of that slow sunrise. Behind me, the dead city has disappeared.

Jack Losh *is a journalist, photographer and filmmaker with a focus on conservation, humanitarian issues and traditional cultures, often in areas of conflict and crisis. His work has been published by the* New York Times, National Geographic, The Guardian, *the* Washington Post *and* Newsweek *among other leading outlets, and shortlisted at the Amnesty, Bayeux Calvados, One World, RTS and Kurt Schork awards.*

A BRIEF ENCOUNTER WITH ALTITUDE SICKNESS

Matt Dawson

Commended 2016 'A Brief Encounter'

It's one in the morning and dark, so I don't see the vomit strike. I hear it, though, a soft patter against my waterproof trousers. My head-torch reveals a lady being helped to her feet by a Tanzanian guide and another man, probably her boyfriend. The couple are late twenties, fit-looking. They sport matching red-and-white North Face jackets.

'You okay?' the boyfriend asks.

'I'm fine,' she says.

'Honey…'

'Don't.'

'We can go back.'

'I can do this.'

Silence lingers between them. The girlfriend's face is devoid of emotion, yet somehow pleads for a warm bed. The boyfriend observes her, plainly mulling over his next words, choosing them carefully. He goes to speak but is interrupted.

'Listen,' she says, 'we're getting to the top, even if I have to crawl.'

I'd read a lot about altitude sickness before attempting to climb Kilimanjaro – the headaches and cough; the nausea and vomiting; the wobbling and confusion; the death – but, until now, I'd never encountered it. Fifteen of us, ten hikers and five guides, had set off from a 4,700m-high basecamp at midnight, all concealing headaches and a muted fear that this invisible killer is stalking us. Behind our concerned faces is a relief that someone else was the first victim.

Our head guide beckons us on, 'Keep going. *Polay, polay.*' *Slowly, slowly.* 'Just two hours to the top.'

We trudge on up the volcano with the body language of prisoners on a forced march. Our progress is monitored by a full moon that mocks our ascent by partially exposing our goal – the dark outline of the crater's rim. Above and below us hundreds of dots of light, in groups as big as twenty, zigzag like plodding caterpillars up the scree slope. Every so often a light drops, the caterpillar pauses, then minutes later sets off again a segment or two shorter. All around people disappear into the darkness as the altitude reaches out for them. It seems totally random. Young or old. Fit or fat. I fall into phases of complete ignorance, noticing no-one, then complete attention as I count those we pass.

Someone asks our guide how far the top is. He says, two hours.

We stop for water near a small cave crafted by falling boulders. Here I observe three types of face: the concerned; the shocked; the washed out.

'You're not feeling it, are you?'

Standing in front of me is the lady in the North Face jacket, one hand on her hip, the other clutching a plastic bottle. 'Got a headache, nothing else,' she adds.

'Lucky,' I say, unable to compute her remarkable transformation.

After the cave the trek takes on a whole different angle, literally. It's steeper, the scree deeper. The groups are beginning to bunch up. Hikers are pausing because they are tired, because they need to eat or drink, because they are dizzy or vomiting. We pass one man who is weeping, pleading to go back down.

I ask our guide how far the top is. He says, two hours.

Soon we are scrambling over large boulders, our walking poles a hindrance as we attempt to haul ourselves up. It's still dark but looking back towards basecamp reveals first light nibbling at the tip Mawenzi, Kilimanjaro's second peak. The volcano's rim now looks unbearably close. Dots of light disappear over the edge as though carried on a factory conveyor belt.

Someone asks our guide how far the top is. He says, two hours.

Ten minutes later we clamber on to Gilman's Point, signalling our arrival at the crater. It's packed, like a platform at rush hour, but we find space near the inside of the volcano where a lake of darkness obstructs any view. We are now at nearly 5,700m and breathing air with 50% less oxygen than at sea level.

'Move, get out the way.'

A man shoves us aside and lays a woman down. I've seen her before but can't place where. By her side is a man grasping a red-and-white North Face jacket. Her face is pale, her eyes unresponsive. A debate ensues as to whether she should continue. This is not the summit, reaching Uhuru Peak involves a further two-hour hike around the rim, they would have to carry her there.

Before any agreement is made, the sun steals our attention away, announcing itself via a prismatic band of light. It unveils a layer of cloud spread out like a protective quilt cast over Africa. We stand mesmerised.

'Please, the cold. Let's go,' our guide says.

As we start to move away I wobble, so the guide grips my shoulder. 'Can't believe I feel fine,' I say, 'I'm not even feeling the cold.'

Matt Dawson *is a writer, photographer and neuroscientist living in London. In his spare time he sits in coffee shops dreaming up new adventures* *or writing about the last one. In 2015, he was a finalist in the Bradt and* Independent on Sunday *Travel-writing Competition.*

ONLY IN INDIA

Deborah Gray

Longlisted 2020 'And That's When It Happened'

It is a truism that India is a land of extremes. On one side of the street, in the savage sunlight, blues are bluer, pinks are pinker; on the other side those same colours are muted by grime and exhaust. The noisy bustle of the marketplace is balanced out by the sanctuary of the temple, quiet as the motes of dust that hang in the air. As a visitor, these are the juxtapositions that charm and intrigue me. It is a country where I can find myself oscillating from immense joy to immeasurable sadness in an instant.

I felt exhilarated driving around Mysore in our waspish tuk-tuk. Its little engine screamed with each violent gear change and we cheated death, somehow squeezing between a decrepit bus and a brightly painted truck laden with watermelons. A brassy Ganesha charm hanging from the mirror danced violently, protecting us as we swerved into the oncoming traffic on an insanely packed roundabout. Clinging on tightly, we rounded a corner and came to an abrupt stop in an unremarkable back street.

It was good to be alive. The blood coursed through my veins causing me to laugh in delight. Springing from the tuk-tuk, I stepped back to record the journey in a photograph – and then it happened, I plummeted through a deep crack in the concrete at the side of the road into the abyss.

Now an Indian drain is not a Western drain that carries away excess water. Oh no, no. An Indian drain is full of black slime; an oozing stream of fetid sludge. Think of a putrid camembert lurking in the back of the fridge; this was worse. Hell hath no stench to compare with that which I encountered seeping from the darkness below.

Years of yoga paid off and I found myself holding a modified warrior pose that allowed me some dignity. The trembling tour guide hauled me out of the pit, his brow uncreasing as I emerged more or less unscathed, but with my full-length, coral-pink skirt coated with something resembling sump oil. The noxious liquid dripped into a mandala around my tarred feet. I was mortified.

Feigning insouciance, I limped behind our guide until we stopped in front of a time-worn door. As it opened, the scent of sandalwood and jasmine wafted past me on a cloud of fine smoke, smothering the rankness in my nostrils; the glory of it had the aura of Ganga's shrine. We had arrived at a perfume factory – of course we had.

I was handed over to a delicate girl with clear mahogany eyes; she could not have been more than 14 years old. She wore a cerise nylon sari embroidered with white flowers studded with sequins; one of those seen in heaps in the market. Her usual job was rolling joss sticks, but today she given a less fragrant task. She led me to a tap embedded in the courtyard wall and began to wash me down. Her dignity humbled me. At first she was gentle, but then she took a huge bar of homemade soap and rubbed my legs, my feet, my skirt – as if kneading dough for chapattis. She breathed rapidly from the effort and her own clothes become soaked. The young girl avoided looking me in the eye, but her shy smile never left her lips. In time the water ran clear and I was cleansed. She stood me in front of the rotating fan in their workroom to dry.

The factory owner then took me by the elbow and led me into a room lit only by the shards of sunlight penetrating the slatted shutters.

He chose a vial from one of the many on the shelves, opened it and rubbed the deep cut on the heel of my hand with lotus oil. 'It is an antiseptic,' he declared proudly, binding the cut while instructing me not to wash it for two days. 'I am pleased to help you. You are welcome here'. Before bidding me goodbye, he selected further flasks of oils and pushed them towards me, telling me to use them liberally.

I entered this building smelling of shit and came out smelling of roses, lotus flower, jacaranda, neroli and waterlily. Only in India.

After a career spent editing and writing cookery books, ***Deborah Gray*** *has broadened her horizons to write about what she loves doing most – travelling. Since entering the competition, she has completed an MA in Travel and Nature Writing at Bath Spa University. Her home is in Malmesbury, Wiltshire, a place she loves for its history and vibrant community life – characteristics she seeks out on her travels.*

THE TAXI DRIVER

Sarah Stewart

Longlisted 2011 'Up the Creek'

I've only felt real fear once, in all my travels alone. The kind of fear that clenches your stomach, halts your breath in your throat, and shakes every fibre in your body awake.

It was the end of a long journey through the Syrian desert, after hours rattling around in a tin can of a minibus, a dozen dishevelled workers sardined into the back. Despite their unwavering stares at the Western woman travelling solo, the interminable dusty road had put me to sleep.

I emerged bleary-eyed in the city of Homs, dumped into the usual melee of taxi drivers vying for business in staccato Arabic and broken English. One young man pounced on my rucksack, threw it into the back of a car that had seen more roadworthy days, and bundled me into the front.

'Keef halak?' the driver asked. Was I OK? I was proud that I'd picked up enough Arabic to respond, and as we joined a current of cars jostling for highway space, I began to tell him as much about myself as I could in his language: 'Ismi Sarah. Ana men New Zealand.'

The alarm bells only rang when, struggling with my seat belt, his hand brushed across my chest as he tried to help. I froze. It was an invasion of space in any culture – but completely forbidden behaviour in this strict Muslim country.

He looked at me eagerly, a typical Syrian youth, over-oiled black hair and a greasy smile. It struck me that I'd been way too familiar with him. In a country where most women don't step out of their homes unaccompanied, where they don't socialise with men, where they cover their heads, I had let down my guard. I was stupidly sitting in the front seat, chatting too freely, to a man who now thought I was fair game.

His arm brushed mine again. I slapped it away. Trapped in his taxi, I was totally exposed – up the creek, and flailing. My skin burned with a torrent of shame at the position I'd put myself in. In a few innocent minutes I'd reinforced every misconception this Syrian man clearly held about Western women.

I had no idea where I was. We were picking up speed through roads that felt totally foreign: past men with red-and-white checked keffiyehs fixed to their heads, past mosques with minarets bellowing out a staticky call to prayer, past children sandwiched between their parents on overloaded motorbikes. I frantically checked the map, aware he could be taking me anywhere.

Finally the Karnak bus station came into sight. Before he'd slowed to a stop I burst from the cab and yanked my pack from the back. A Syrian policeman, his face contorted in anger at seeing me emerge from the front, began to berate the driver. I hung my head and left the sleazy man to his fate.

But I wanted to be anywhere but here: yet another chaotic bus station, hundreds of buses parked haphazardly, their destinations signed in swirling Arabic script which I couldn't read. Exotic place names tripped off the drivers' tongues, 'Aleppo! Aleppo! Palmyra! Palmyra!' Wandering though dust and rubbish, the smell of rotting fruit hanging in the oppressively hot air, I couldn't find the bus to Hama anywhere.

I didn't want to make contact with anyone else in this sea of men. But, fighting back tears, I approached the most respectable-looking man I could find, and asked for help.

Khalid Al-Hamsi was my paddle – a 42-year-old university lecturer from Hama – and every ounce the chivalrous man that his dress implied. He steered me to the ticket office, and on to the same bus as him. 'Are you having a good time in my country?' he asked, concerned. I lied.

But as we took the road to Hama, I slowly relaxed. After days in the desert, it was refreshing to see shoots of green. Olive trees stood out against red soil. We crossed the Orontes River, carving a path through a deep gorge where its water feeds fertile fields of wheat and barley. Khalid fed me his wife's date biscuits, and showed me photos of their wedding.

Here was the Syria I had come to love, a place of genuine warmth and kindness. As the bus pulled in by the waterwheels at Hama, Khalid wrote down his phone number and told me if I was ever in trouble, I was to call.

I had learnt my lesson. I had learnt never to forget the culture I was travelling in. But, more importantly, I had learnt that in those rare times when you're really up the creek, a paddle will always drift by. You just need to be open enough to reach out and grab it.

Sarah Stewart *is a New Zealand journalist who spent seven years exploring the world, often solo. She is grateful to have experienced life in Syria before war broke out. She now lives in Auckland, working for a charity tackling child poverty. She's delighted that her two young daughters have inherited her love of travel. She blogs at cessandasuitcase.com.*

WRESTLING WITH RED TAPE

Steven Tizzard

Longlisted 2014 'Meeting the Challenge'

Sitting in an internet café in Almaty popular with gamers, I'm one click away from purchasing two flights to Istanbul, bypassing Uzbekistan, Turkmenistan and Iran. The end is near. The cursor hovers. A few hours earlier the squat Uzbek official with thinning, sideways hair returns from a smoke break. Behind the glass screen in the dingy consular section, he produces our passports from a drawer. Pointing at my British passport, he says, 'Yes. OK.'

My girlfriend's New Zealand passport, he waves, shaking his head, and slaps it on the counter. He walks out as if insulted. I slump in an armchair presuming he's rewarding these few seconds' work with a further cigarette.

We have two days to get out of Kazakhstan. Zero chance of permission to remain. The road to Uzbekistan is blocked by red tape. Turkey is our most realistic option to continue the journey west. But there is an alternative. Do we have the energy for it? Enough time?

A fortnight of limbo already endured in Almaty, where the overland backpacker is tantamount to vagrant. The city doesn't seem to like us much and it is difficult to feel any affection back. The hospitality of the people has been left far out on the steppes with their old nomadic ways. Not that it matters. We're just passing through.

The January daylight is short, but the days and nights linger. We are in a routine that resembles something from an M.C. Escher print, dragging our feet through the city's grey snow.

The daily humdrum is interrupted by visits to the Embassy of Uzbekistan on a slippery street on the edge of downtown. Our first vigil outside is on a finger- and foot-numbing -13 degrees Celsius afternoon. We achieve nothing more than getting our names on the waiting list.

A few days later we return in the morning only to be told we cannot add our names to the list until the afternoon. We return after lunch. A group of Afghan men crowd around us, asking questions. We remember each other from before. Nudging me in the ribs, one says, 'Someone died here last year. Waiting. In the cold.'

Another tells me he was deported from Greece and Turkey for dealing in 'contraband'. He asks me to write him a letter of recommendation for a UK visa. Sensing my reticence, he adds, 'I have money and I can be useful for you. I speak nine languages.'

The Kazakh guards have a hut outside the gate of the embassy. It's where the deals are done. An elephantine, ageing American, working in the mining industry, arrives late, but doesn't wait. Many others jump the queue.

Again, we don't get beyond the waiting list. It isn't until the fourth attempt, near to closing time over four sub-zero hours later, that my girlfriend is admitted with our passports. Everything appears in order. Time is running out yet we still have time to kill.

Less than a week later, deciphering a conversation with our hotel receptionist about a phone call, we realise we've been summoned. The ice is melting, dripping from the evergreens in Panfilov Park around the sugar-icing-coated Zenkov Cathedral. We go early, get our name on the list, and wait. The guards are mingling, sharing cigarettes. Everyone agrees things are warming up. A little over an hour later our names are called. A guard unlocks the gate and escorts us along an alley to the office.

After the consul's abrupt departure, his aide comes in: a young woman with bright lipstick, not yet hardened by sob stories or a demanding boss. 'I'm sorry. You do need the letter of introduction. It is Tashkent's decision.'

'But you said it would be OK,' my girlfriend says. 'That I didn't need a letter.'

'I know. I'm truly sorry. There is nothing we can do.'

I sense she is genuine. A messenger put in an awkward position. My anger dissipates.

'You could,' she says, 'go to Kyrgyzstan. Getting visa is more easy.'

'Kyrgyzstan,' I say, slouching in the armchair.

'Yes,' says the woman. She passes a business card of a travel agent to my girlfriend. 'They can help you. If you need.'

We leave in silence, led out past the consular official, who lights another cigarette, looks away. Outside, we wave to our Afghan friends and trudge to an internet café, decisions to make.

I close the web browser. We go the next morning to the Consulate-General of Kyrgyzstan, a suburban chalet not far from where the cable-car begins its ascent up *Kok Tobe* (Green Hill). Later the same day we return to collect our freshly stamped passports.

The following morning we're riding a *marshrutka* (minibus taxi) south to the capital city of a country I can't even spell correctly.

Having spent a year on a kibbutz, ***Steven Tizzard*** *got dysentery in Damascus and resolved to quit backpacking. The vow lasted less than a* *month. Since then he has roamed around, written about and taught in South America, Europe and Asia. Where possible, plans and planes are avoided. He and his wife, a Kiwi, have two spirited sons born in Japan.*

NUPTIALS

John Wilcox

Winner 2004 'We've Come a Long Way'

I'm a soft tourist – a very soft tourist. I don't do canvas. I don't do rough. So why was I drifting down the Amazon with a wounded guide and no map? Perhaps we should begin at the beginning...

The wedding was in Rio. I agreed readily to attend an old friend's nuptials on the other side of the world. But I surprised myself when I also signed up for a two-week tour, including four nights in the heart of the Amazon.

The wedding weekend was a gem. The boy from Sheffield tied the knot with the girl from Ipanema. After the hangover, two days of essential sightseeing – by taxi of course – before packing my bags for the great adventure.

Two hours away from Rio by plane is Iguassu Falls. It makes Niagara look like a garden tap. Enough said.

Onward nearly a thousand miles to the northeast next and the colonial charms of Salvador. Narrow twisting streets shelter countless churches and fine but faded colonial buildings. I gorged on spicy food from Bahia and danced samba through the night.

So much for the warm-up games. Now for the big one. Soft tourist against mighty river.

First, an unexpected shock. The Amazon is beautiful. The landscape is breath-taking. So I completely forgot that I don't do boats either. But on the way to our river base I negotiated four of them – each one tinier than the last.

The Amazon Lodge is the pioneer of Brazil's green tourism. Floating around a central island, four spines stretch out like the spokes of a wheel. Three of these have tin-roofed bedrooms. The fourth has four WCs and four cold showers. The 'restaurant' had zero-rated ambience and décor. But it actually served the best food I'd eaten in Brazil.

When I fell exhausted into my minuscule tin-roofed room with equally tiny bed, a sink and not much else, my soft shell was in danger of cracking completely. But then the magic began. I awoke to a sunrise in heaven. The water is still. The canopy of trees would inspire Constable and the only sounds are made by thousands of unseen monkeys. Eat your heart out, Attenborough. This exploration thing is getting into my blood.

Did I say blood? Of course, that's where we came in. Activities at the lodge were well organised. Jungle tour in the morning. Waterside village in the afternoon. But the casualty list mounted. Kevin – gippy tummy. Thomas – sore foot. Monika – sunburn.

Now, I don't do fishing. But someone had to go. So the piranha expedition consisted of me and fellow wedding guest, Daniel, plus the boatman and our very handy-looking Indian guide.

Our guide was very keen on education. His self-taught English was pretty good. He had ambition and charm and he had come a long way from his beginnings in an Amazon village. But that's another story. There's fishing to do here.

After a short lecture on piranhas, with assurances that they only killed people in movies, guide and boatman set to with rudimentary fishing kit. Within minutes, the boatman hauled one in. Even Daniel hooked one. This was like shelling peas.

To prove a point about piranha teeth, our guide picked one of the finny victims from the bottom of the boat. He told us once again how they never attacked humans. These were nearly his last words. The second captive had had enough of seeing his cousin tortured for education. In a last defiant act, he leapt fully five feet from the bottom of the boat and lodged his gnashers deep in our lecturer's index finger.

The blood began to fountain out. The fun fishing trip had suddenly become a darker and more dangerous thing altogether. The guide was in shock. The boat was an hour away from base. The boatman didn't speak English. There was no first-aid kit. Suddenly, I could see the value of health and safety officers.

Now as it happens, Daniel had been a boy scout. So, when he produced his hanky I followed his orders. I gripped the dangling finger and held it tight while Daniel applied the makeshift tourniquet. We dripped blood all the way home.

This story of adventure won the Bradt Travel-writing Competition in 2004. ***John Wilcox****'s prize took him to Bosnia and after that trip, he and his wife set up a small charity, FairPlay, to raise money to build a safe children's playground in the aftermath of the war. He has also contributed to past editions of the Bradt guide to* Bosnia & Herzegovina.

TAK JADI

Jude Marwa

Longlisted 2013 'A Narrow Escape'

Walking out of arrivals I winced as the damp, warm air of KL engulfed me. I made a bee-line for the first taxi and dived into the back seat. Ignoring my usual ambivalence, I relished the crisp coolness that the air con rescued me with.

'Desapark City, Kepong please… on meter.'

'Yes mam.'

'Your first time here mam?'

'No I live here, been here two years now.'

'You have children?'

'Yeah I've got two, and you?'

'Yes mam two boys, 4 years, 6 years.'

'Ah, bet they keep you busy!'

'Oh yes mam they very busy yes mam, but wife she wants girl, you… boys?'

'I have one girl and one boy.'

'Oh lucky, yes?'

'Sometimes… sometimes not!!' I laughed.

Relaxed, I closed my eyes and rested my head against the cool glass of the window. Sometime later I awoke and looked around for the familiar landmark of the PETRONAS Towers. There they were – behind me? That wasn't right. I was confused. I leant forward so the driver could hear me over the drone of the engine.

'This isn't the route I usually take.'

No answer. I raised my voice.

'I haven't been this way before, you sure you know the way?'

'Yes mam… this different way.'

The driver looked back at me in his mirror. Our eyes met and his darted away.

I started to feel uneasy.

'This doesn't lead to Desapark, can you turn round please, this is wrong way.'

There was no response. I glared in the mirror willing him to look back at me, but instead he stared ahead, stony faced.

I felt panic racing through my body.

Looking around the taxi for some peace of mind I noticed his ID taped to the dashboard. I grappled with my ego momentarily before taking my phone out to call my husband. Of all the times to hear those words, 'you have insufficient credit to make this call.' I was about to hang up when I realised my only hope was to pretend. So I spoke into the dead line.

'Hi, yes I'm in a cab but he's not taking me home. I don't know where he is taking me. His name is Lee Hong and his number is RK522340. Yeah, you can track me on your iPhone. Exactly, if you're already in the area you'll catch up with me in no time. Yeah I'll stay on the phone.'

I clocked the driver looking at me again. I hardened my gaze. We turned off the highway as he drove at speed into a village and took another left. We were getting further away from the bustling streets. We were heading towards jungle. At the next junction I tried the handle on the door, ready to make a run for it, but it wouldn't budge.

We took a final bend and I saw ahead what I had feared but not allowed myself to fully believe. Four men stood at the roadside facing the taxi. We began to slow. My heart raced. I gave one last pointless piece of information to my fictional rescuer on the phone.

'I think he's delivering me to some men. I think I'm being sold.'

I felt a tear roll down my face. This was it. The end of everything I knew, my life about to become something people read about.

The taxi pulled up, the engine still running. The driver's hands shook as he fumbled around in his cubby. He pulled out an envelope. I felt my chest rise, my throat tighten. More tears fell. The car echoed with the sound of my breath. The driver handed the envelope to the men.

'Tak jadi! Tak jadi!'

I watched him wipe the beads of sweat from his eyes as he passed an envelope through the window of his battered old Proton.

'Tak jadi…' his voice broke.

The men looked at each other. One stepped forward and booted the driver's door with his foot.

I let out a yelp.

The taxi driver cowered, and began to ramble, desperation oozing out of every word. Another of the group stepped forward and grabbed the driver's face.

Then I heard a word I understood.

'Tracker.'

The men stopped shouting and looked at me.

My phone was in my hand.

Before I knew what was happening the envelope had been snatched, the gang had mounted their bikes and they were gone.

Weeks later at work I heard a colleague say say, 'Tak jadi.'

My blood went cold as those words took me straight back to the taxi.

'What does that mean ... tak jadi?' I asked.

'Oh,' replied my colleague, 'It means cannot happen, won't work out.'

Jude Marwa *combines her chaotic wanderlust with a love of storytelling and has been lucky enough to have had two stories published by Bradt. Jude is the founder of Chirpy Bakers, which produces delicious brownies made from insect flour. Find out more about how to feed the world without starving the planet at* *chirpybakers.*

A STRANGER'S SMILE

Zoe Efstathiou

Finalist 2012 'A Close Encounter'

'I'll see you soon,' he says, squeezing my hand. In India, hand-squeezing has become our way of kissing in public. I should just go and catch my train but we've had such a nice day together. He quickly glances over his shoulder to see if anyone is looking and then leans forwards and plants a kiss on my lips.

'See you soon,' I say finally as I let go of his hand and swipe my token across the barrier of the Delhi metro station. He winks and walks away.

I am halfway down the escalator when I can feel someone looking at me. I glance over my shoulder and my eyes meet with those of a man in his fifties or sixties standing behind me. I am wearing a vest top and a pashmina. Like a reflex, I adjust my pashmina to make sure my shoulders and chest are covered. I hear the man mutter something under his breath. He steps forward to stand right next to me and turns his head to face mine. He repeats the word, but I can't make it out. A slur in Hindi, probably.

His brown eyes are narrowed with a look of disdain. The skin around them is lined with dozens of fine wrinkles. I fiddle with the zip on my handbag. *Don't be intimidated*, I think. *Ignore him.* I look ahead, but I can see his face out of the corner of my eye, staring at me. *Ignore him,* I think but then I quickly glance over and see that yes, he is still staring at me. There is a black strap across his chest. I see that it holds a small, curved knife, kept in a red painted holster at his hip. *What the hell?* I feel something burn through me, a flash of fear that quickly turns to anger. Anger at being intimidated for no reason, anger that a man at least thirty years older than me with a menacing little knife feels he has the right to come and stand next to me and mutter under his breath. *Who does he think he is?* I look right back at him. *You don't scare me,* I think and we stare at each other until the escalator reaches the ground floor. I look away and smile to myself just to make it clear that he has had no effect on me. I walk to my platform, not bothering to look back.

I look at my watch while I wait for the train. It's 10.15pm. The hotel I'm staying at has an 11pm curfew but it's only three metro stops away. I watch the train approaching and then see him, the man with the angry eyes, walk on to the platform, his hands in his pockets, glaring at me.

We both get on to the train through doors a few metres apart. Keen to get away from him, I walk through at least six carriages. I eventually stop and pretend to study the train map above the carriage doors before glancing back, only to see him standing further down the carriage, staring at me. The knife seems bigger than before and the carriage is empty. I feel the electric mix of fear and anger and march down the train as it leaves the platform and starts pulsing through the veins of the underground.

Why is this train so empty? I wonder, panicking. I glance over my shoulder and see him striding after me. He is no longer looking at me and wears a blasé expression as if what he is doing is completely normal and non-threatening. *What does he do with that knife?* I think and imagine him cornering me, dragging the curved blade across my face. *Oh God. Oh God,* I plead as I approach the end of the train. *Where do I go now?*

An automated woman's voice is saying something. I realise she is announcing our arrival at the next stop. The train pulls into the station and I leap off and practically run down the deserted platform. I look back and see him walking briskly after me, his knife flapping at his side.

I jump back on to the train to confuse him but he follows. He is standing at the end of the carriage and casually loops his hand into an overhead handle. He looks right at me.

The doors begin to close. My heart is racing. I make a dash for it and throw myself through narrowing gap. I scan the platform. He's not there. The doors are closed. The train is pulling away. And then through the window I see him and he smiles.

Zoe Efstathiou *is an author from Oxford. She writes romantic comedies under the pen name Zoe May, published by HarperCollins. Zoe also writes thrillers with her debut thriller coming out in 2022. When Zoe is not writing, she enjoys reading, cooking and painting. Zoe likes to travel whenever possible and has been most inspired by her trips to India.*

6
LESSONS & LIFE

"I stared until my head spun with the vastness of it, and understood."
Catriona Rainsford

Here is writing that shares learning: new skills; new knowledge; new insight. We are not the same after the trip.

Uganda Kosovo Hungary Andaman Sea Turkmenistan China India USA Kenya Mauritania Ethiopia Peru Kenya Syria Chad

CLOSER TO HOME

Kirstin Zhang

Winner 2020 And That's When It Happened'

I had swum too close to the sun. The day before we'd gone down to Lake Nabugabo at the Equator to escape the capital which was still jittery after the explosion of three bombs and nearly eighty dead. Now I lay on a day bed in the deep shadow of a friend's veranda hooked up to a drip. Her bungalow lay within the compound of an international AIDS hospital, and one of the doctors had come down and diagnosed heatstroke.

I lay in half-sleep, my reverie broken only by the *thwack thwack* of the gardener's machete as he cut down jackfruit, and the regular patrol by one of the compound guards, who strode across the lawn, an AK-47 slung casually over his shoulder. The imminent arrival of Colonel Gaddafi for a meeting of the African Union had heightened the tension. Despite the billboard messages all along the highway between Entebbe airport and Kampala welcoming the African delegations to Uganda, Gaddafi was no friend of the Ugandan president, who was suspicious of the mosques he funded in the northern Ugandan provinces and his vision of a United African presidency. Already rumours swirled that Gaddafi was complicit in the recent atrocity... that he was trying to destabalise the country as presidential elections loomed.

For a day or two our planned trip to Jinja seemed unlikely. I'd had my heart set on visiting the source of the Nile and the spot where some of Mahatma Gandhi's ashes were scattered. But by the third day the nausea had settled and I could stand. The back seat of my friend's car was filled with cushions and ice packs, and a bucket. Just in case.

Jinja was cooler. Mist crept up the banks of the Nile and swathed the gardens surrounding Gandhi's memorial. Our faces were rain-streaked as we offered our strands of marigolds. Afterwards, we took shelter beneath the town colonnades which were filled with cafés and fabric shops. In one I was struck by the Birmingham burr of the owner. They were ethnic Indians, they told me, who'd fled in 1973 after Idi Amin's expulsion of the Asian community. They'd travelled first to Tanzania and then on to the UK. A new law now allowed them to reclaim their property, and they'd returned after nearly forty years. Their daughter, Rita, however, had stayed on in Birmingham. 'She's got a family now, and no longer thinks of this as home.'

And then it happened. As it had throughout this trip. This time it was a word – 'home'. Previously it had been the smell of green bananas roasting over a roadside fire, the sight of a hillside, the red earth turned over and ready for planting, and the sound at night of fruit bats in the pawpaw tree outside my bedroom window. I felt the deep ache of something lost. I'd grown up in the 1970s in Papua New Guinea, in the wetlands, among the abandoned airfields and ammunition dumps of the Pacific War. When we weren't scavenging for spent cartridges, my best friend Sangeeta Shrivastava and I would lie beneath her parents' bed reading her father's *Playboy* magazines, while she told me hair-raising tales of Idi Amin. They too had been refugees from Uganda. But the place where I'd spent my first twelve years was also descending into post-independence turmoil. One Saturday, returning home from swimming lessons, I was told we were leaving. Thirty-five people were dead after demonstrations, and the Prime Minister had declared a state of emergency. That night I left my home, my dog, and my father who was bound to see out his contract. I never had a chance to say goodbye to Sangeeta.

The rain lashed our windscreen as we sped through the Mabira Forest on our way back to Entebbe. We passed groups of men playing checkers on dripping verandas, goats shivering as they searched for weeds among the rubble, and a sign which read 'You were lost and now you are found'.

By the time we reached the outskirts of the city the sun was blazing, but now we were caught up in the rush-hour traffic. As we navigated potholes hidden by surface water, we heard the rotation of helicopter blades, and then, somewhere behind us, sirens.

'Shit,' said our driver Moses, as an armoured car tried to nose ahead of us. It was flying the flags of Libya.

'Gadaffi,' whispered my friend.

There was the horrible sound of metal on metal. An army jeep forced us up on to the sloped embankment. The six tall soldiers kneeling on the back, Kalashnikovs in hand, didn't even glance our way.

I began to sob.

'Please auntie,' said Moses, 'don't be afraid.'

But how could I explain to him it wasn't fear that made me weep?

Raised in Cyprus and Papua New Guinea, ***Kirstin Zhang****'s interest in people and places was piqued young; the family often travelled with her father during extended business trips around the Commonwealth. Following three years in Japan, she studied Japanese politics and anthropology at the School of Oriental and African Studies in London. She works for the national arts and screen agency for Scotland.*

THE VILLAGE SLEDGE RUN

Alan Packer

Winner 2017 'Lost in Translation'

The first annual Kosovo snows are met with dismay by the parents; exhilaration by the kids. Nazmi frets about freezing pipes when the electricity is cut and the circulation pump stops. He recalculates the cubic metres of wood stacked in the yard, assessing whether the early winter will exhaust his store, 'If I have to buy more later it is double the price.' The early bird catches the dry wood and enough of it. He has often tried to explain wood to me but I never quite get it. I am certain of inconsistencies in local names and suspect the same wood has a different name each time I ask. Nazmi explains, 'This burns fast, it is good to start. This burns hotter and we use it more on cold days. This is good at night, it is long and slow.' These could be 'ahu', 'bung' or 'qarr', today at least; the texture and weight are the real language. In regular Albanian everyone is happy that rabbit and hare are the same, like tortoise and turtle, mouse and rat; bird is bird, mainly.

I creak open the oak door from the yard and snow powders me. The children have been sledging since early and the runs have crushed the snow into ice. Armend shouts, 'Watch, watch!' and redoubles his determination that this run should be fast and long. I watch with some admiration as he flies past the neighbour's crumbling stone gateway, scrapes his shoe into the snow to brake and turn, negotiates the unguarded bridge over the stream and heads on down to the school.

I slither along after him, dodging Arber, Anita, Arijeta and Albion as they dare the same route. It always strikes me as odd how many names start with A, like Albania, as if the imagination stretch to think of B or Z is too exhausting.

I congratulate the children as they gather themselves for the trudge up for the repeat. I test out my Albanian, piecing together 'rruga' for road, 'plot' for full and 'akull… something' for ice, and shout, 'Rrruga ështē plot me akullore.' Suddenly the whole group of village boys are overcome with laughter. The girls, a little too embarrassed to laugh, glance at me and turn away in explosions of giggles. Their mirth subsides and they catch each other's eyes and smile. Then, more excited by the snow, they all yank on their sledge tethers and head up the hill.

I go off to find Nazmi and hope to be invited for tea. He is satisfied with his inspection of the wood, water and roof. He calls, 'Hajde, qaj.' We head out of the cold. Strong tea is poured into the glass and hot water tops it up, a chunk of lemon. The lemon is a sop to the foreigner and substitute for my refusal to take three spoons of sugar.

Armend and Anita arrive, their boots kicked off outside and the glow of cold on their cheeks. They politely shake my hand and wish me a good day as is customary, a questioning shyness playing around their continuing amusement.

Armend whispers something to his dad. Nazmi slaps his knees and gives me a friendly thump on the shoulder. Whatever I said down the hill is an additional highlight to the morning's sledging. Unable to ask for an explanation for every odd thing that I fail to understand, I let the friendship carry me through the unknowing. Daring to trust takes courage. I sip the tea. I smile. They smile. Anita giggles. I laugh.

'Okay,' I say, 'What, what?'

Armend finally breaks his barrier of polite respect and, grinning, says, 'You said the road was full of ice-cream!'

Alan Packer *advises on the development of local government in the Balkans. He began travel writing inspired by vivid experiences of living in Kosovo. His competition story 'The Village Sledge Run' is based around 'Hangjik', a traditional village property that he helped to renovate. The current focus of his writing is to link historical literature with modern travel. From Yorkshire roots, Alan's UK home is in the Scottish Borders.*

SUN AT PEACE

Liam Hodgkinson
Winner 2015 'Serendipity'

The train takes its final breath in Central Transdanubia, in a valley south of Slovakia, as the grey and tired eyes of dusk close. I shouldn't be there. I shouldn't hear the engine's final rattle, that CLUNK CLUNK SLAM followed by silence. I shouldn't see black smoke rising as the world stops going east – the low-built houses on the horizon slowing, slowing, and then finally still. I shouldn't stand up, choke, and follow a pale uniform out into the Hungarian October, shake off the last 14 hours like a coat, and watch the sky over Tata, Vértestolna and Tardos bleed a rose-garden red.

I should be 35,000 feet higher, observing this unfold from a point above, passing through the invisible corridors that stretch from cloud to cloud between London and Budapest. I should be sipping dull tea with a stranger's knees in my back. I should be buckling and unbuckling my belt to the call of a red light. But I don't trust the laws of aerodynamics. The air vibrates. The wings bounce. My mind stampedes. I know fear up there, and this train is how I've talked myself out of it.

So my day begins before the sun's. Platform nine. Brussels Midi. I breathe the cold in silence and watch the numbers flip on the departure board until they become the ones I'm looking for. The train door opens with a morning groan, and the conductor steps down and tips his hat. En voiture, he says. All aboard. I find an empty carriage and wait. The engine stirs, clears its lungs, and as the first train to

Frankfurt departs, I rest my head against the window and gently rock alone towards dawn.

I'm out until the iron bridge spanning the Rhine jabs me awake. Belgium has passed in a sleep and the clouds have collected until there's no sky, just an empty colour saying: today it will rain. Outside is Cologne and a medieval cathedral with spires that pierce the weather like teeth. The city's streets are busy with trams, and the trees that line them blaze a golden orange. The leaves have started to fall and as the train picks up speed I watch them fill the gutters like confetti.

The hours tick on and the woman to my right gets older, then younger, turns into a child and later a middle-aged man. I move through Germany to Munich, past the stone towers and red roofs of Nuremburg, and the foothills of the Alps lined with stoic pines and green water so clear I spot scales swimming through it. Somewhere I pass into Austria, and the pastures grow so thick and lush I think of Anschluss, Do-Re-Mi and the Von Trapps dancing in old Hollywood.

Now and then, this nostalgia is cracked by darkness. The air compresses, my ears pop, the world disappears. I hold my breath in these tunnels. It's involuntary. I don't know why. And when the land and earth come piercing back, the hills seem brighter than before, and I ask myself, am I dreaming, or have I now, this second, woken up?

The motion stops not long past Győr, at the meeting of the Mosoni-Danube and Rába, in Hungary's Little Plain, on the final straight to Budapest. I smell the burning before I see it. Like toast left unattended or a neighbour's bonfire. I'm tired and want to be moving, and it's not until smoke clogs the carriage that I accept my journey has ended.

Minutes later I'm sitting on a straw-yellow ridge as the engine steams like a kettle. A man in a dirty workshirt offers me a cigarette.

'Very sick,' he says, pointing at the train.

'Me... you... going nowhere.'

I haven't smoked for years, but he's right, so I take two and wait.

Then, suddenly, the day is over. The light a poem; this its final stanza. The evening gathers in the branches of the oaks, and the cars in the distance are going home. Against the sunset, an old farmhouse pours purple shadows across fields of wild flowers. A plane roars overhead, marking the sky. I stand for one long minute and watch it leave. When I started, I thought today was lost, but as the night closes in and the hours run together, I know different.

The Hungarian word for sunset is napnyugta. It translates literally as sun at peace. I witness it sink and for a brief moment I'm thankful for my fear, and as the lavender disappears I wonder about this electric feeling, this life with myself. I smoke and watch the light leave the blue-river hills and as the stars begin to talk, I say to them, quiet, not even a whisper, where will you, NO, where will this dread take up and bring me next.

***Liam Hodgkinson** is a writer and editor based in East London.*

MEETING THE CHALLENGE

Lucy Clark

Unpublished Writer Award 2014
'Meeting the Challenge'

I let my body go slack as I am dragged into the darkness. I am pulled, I am sucked, I am punched. My eyes and mouth are shut tight. I am a rag doll casually flipped over and over and over, the least favourite toy of a bored child. I try not to breathe. Then there is a lull and I momentarily regain control but there's a sharp tug on my ankle and I'm off again, dragged forwards, arms flailing. I've let go of my board. Somewhere, I faintly hear a man shout, 'don't let go of your boooaaarrrd!'

And now I'm struggling to my knees in shallow water. I still can't see because my hair has come loose and is plastered over the front of my face. I'm fighting for air while simultaneously pulling up my bikini bottoms and emptying them of sand. And I'm laughing, but there are tears and snot and sea-water pouring down my face, too. It's day three of the surfing holiday and I'm beginning to wonder if this just isn't my sport.

Back in the sky-blue jeep, ten of us sit pressed together, salty knees touching as we bump along the dirt track, discussing our minor triumphs and major setbacks, wondering what breakfast will be. Outside is India: a man in a grubby dhoti stands in front of his corrugated iron kiosk scratching his belly with one hand and brushing his teeth with the other; a family of four on a motorbike overtakes us then narrowly misses a goat, horn sounding loud. A puppy hangs off the teat of its exhausted mother as she ambles towards the shade, four further puppies around her paws. Large, stainless-steel bowls are balanced on the heads of women in immaculate saris of pink, gold and pale blue, and Hindi music blares out briefly as we pass a stall selling bootleg CDs. An elderly man wobbles along on a rusted red bicycle, blue plastic bags hanging from the handlebars, while the temple elephant in chains mournfully looks on. And children, children are everywhere, going to school, playing in the ditch by the side of the road, in the arms of their mothers, and they all wave and smile and shout at the sky-blue jeep piled high with surfboards and foreigners, and the driver beeps his horn and we all wave and smile back. The smell of dosas frying and vegetable curry hits us and we groan with hunger, longing to fill our bellies with something other than the Andaman Sea.

I keep going back to the water. I go back while others have a day off, a lie-in, but I won't because although my body is bruised, scratched and sore, although my shoulders hurt like hell and I'm showing no signs of getting close to standing on my board, the chaos of the water is addictive, it's liberating, it's frightening and it's letting go and living, and beyond the white water, when I get there, is endless blue and peace and silence apart from the occasional plop of a flying fish.

Day seven and the last drive to the beach, where a small community of fishermen live in hip-height huts made from palm trees, their wooden boats tipped upside down on the sand. I glimpse inside one of those huts as I struggle along the beach with my board and see the very barest essentials of life. The waves are gentler today and I manage to paddle out beyond the break without being knocked backwards to the beach, and I sit on the board in the green water and compare bruises with my fellow pupils, chat aimlessly about lives back home, our new curry addiction, the next wave. There's a shout, 'the next one's yours!' and I look behind me to see the green curve coming towards me, frothing at the edges, and there is only now and I begin to paddle hard and I'm moving through the water, I'm not fighting it, it's carrying me, I feel the board lift up and somehow I don't topple backwards this time and the board is moving and I manage to bring my feet to the middle and I'm up, it's a crouch and it's off balance but I'm on my feet on my board and I'm moving through the water and my stomach does a flip and I even manage a smile before falling back into the blue.

A languages degree and a love of being somewhere new took ***Lucy Clark*** *around the world, including spells living in Mexico, Germany and the United States. Unable to cure the travel bug, she made her career with it and now specialises in rail and river experiences for Belmond. She lives in London with her daughter and continues to write as a hobby whenever she can.*

TURKMENISTAN BLUES

Helen Watson

Finalist 2014 'Meeting the Challenge'

The Karakum Desert blisters under a virtual sun on my computer screen. These days my life is all too full of challenges that don't involve breaking a sweat and the figure of myself cycling towards the horizon along a road flanked by sand and scrub is waylaid by a band of desktop icons belonging to documents that I have been too lazy to file: 'Employ_satisfact_survey.doc', 'Conference_reg_Nov2013.pdf' etcetera. Sometimes I think that they are closing ranks to block my former self out of the picture.

The wind blew as a hot wall against us, drying eyeballs and tongue and singeing cheeks smeared with four days of factor 40. Ed, my husband of two years and nine months, was behind me and I, pedalling hard, counted the strokes until I could drop back and tuck into his slipstream. An ache radiated through my legs and the dust-caked teat of my frame-mounted water bottle was a constant reminder to be thirsty. Ed would lead for twenty minutes after my ten – efficiency left no room for pride – and then as the afternoon winds strengthened against us we'd switch round and round for hours. Chains rattling.

In the nine months since our overladen start from Glasgow this journey had forged us into a single machine. When we set up camp now, pegs were placed and dinner cooked without the need for detailed discussion. There were 130 of the 547 kilometres to the border left and only 24 hours on our five-day transit visa – the only sort issued to the unaccompanied traveller by the Turkmen government. We both thought this crossing would be the greatest challenge of the journey to China.

'Time's up,' Ed said. I fell back.

The pain eased and I took a swig from the bottle, washing warm water and grit past my teeth. I focused on the turning tyres and the asphalt disappearing.

We caught only glimpses of this land: further south there had been villages next to canals of brown water lined with rustling poplars. Cotton plants wilted in fields of baked clay. A group of women, waiting for a bus, crowded around gifting us a bag of spiced gingerbread. Their paisley scarves in greens, gold and maroon swept hair back from Turkic cheekbones. In the town of Mary, state buildings of marble and glass rose from pavements swept by old women with bundles of twigs. Schoolgirls walked by in uniforms of emerald silk, the colour of the national flag. Two farmers with labour-worn faces let us camp on their land one night. Sharing no language, we cracked sunflower seeds between our teeth and laughed at their donkey together. Truck stops sold mutton stew and green tea. Once we reached the true desert, cars would occasionally slow next to us, an arm stretching from the windows to offer a swig of water or vodka. Nomads touted plastic bottles of fizzing yoghurt under the fierce sun and camels plodded the verge.

'Time's up,' I said. Ed fell back.

We cycled until the sand dunes were dark silhouettes against a navy sky and set up camp in a hollow. I tore pieces from a wheel of dry bread and soaked it in hot stock for dinner as Ed rolled out camping mats. We drank our ration of water as tea and lay together on a billion grains of sand, watching a billion stars. We held hands, too salt-encrusted and sore of body to entertain more than that. The Uzbek border was 89 kilometres away.

'I think we'll make it,' I said.

Ed kissed my cheek in the darkness.

Beyond lay Uzbekistan. In two nights' time we would be in Bukhara, eating double portions of noodles in a café set under mulberry trees, skin tingling after a scrubbing in the *banya*. Then we'd traverse the plains to Samarkand and curve away to the jagged teeth of the Wakhan, the Pamir and, finally, into China.

Further ahead still, there was returning. There was finding employment in a recession and, even four years later, the adjustment to a stationary routine. Hardest of all, with the parallel tracks of our lives forced increasingly apart by work, there was protecting enough time together for the old adventures.

It's 20.03 in the evening and the Karakum Desert blisters on the computer screen as I close the door to my office. I can taste that grit. I can almost feel the hot wind cracking my lips, but now it's the simplicity of the two of us and the desert road heading out to the horizon that tortures.

This absorbing tale was a finalist in the Bradt Travel-writing Competition in 2014. We have since lost contact with the author.

VANISHING ISLANDS

Michelle Wu

Finalist 2010 'The World at My Feet'

This old lady would be the final try.

With a sigh of resignation, I asked my question again. 'Excuse me, where can I find the old men who practise calligraphy?'

'Wrong gate,' she replied abruptly with the Beijinger's accent of exaggerated 'Rrrr'. My heart sank. Fate had not guided my feet to these elusive calligraphists at the Temple of Heavens Park. My repeated attempts of 'human Googling' around park attendants all week pointed me to contradictory directions – except to the place I was seeking.

The lady continued sweeping twigs with her worn-out broom. Her face was barely visible under her oversized bamboo hat. 'They are at the South Gate. Every morning. From 6.30 to 9,' she said with certainty. 'Want to go there?'

' Yes…'

'Follow me. They should still be here. I am going in that direction for my job anyway. You can find them near the Altar, by the Echo Wall.' I lacked the eloquent Mandarin phrases to show my gratitude, so I flooded our conversation with 'Xie Xie' after every word she uttered.

As we walked across the park, I discovered a hidden side to this urban oasis that awakens only during the morning hours. The park transformed into a promenade theatre of life. Al fresco tango-dancers twisted and turned under one cluster of trees; folk dancers nearby drew pictures in the air with red silky fans; Tai Chi devotees practised mindfully in secluded spots; youngsters rallied in improvised tennis courts where strings tied across tree trunks became the nets. Soon we reached a white pagoda.

'I am leaving you now. Walk beyond this tower, they'll be there.' She walked away before I could thank her.

Will they be there?

As I walked towards the towering red portal frame of the South Gate, I saw a few silhouettes moving serenely near the willows. My feet stopped when I saw a grid of vaporising Chinese phrases on the ground. The world of Chinese calligraphy was beneath my feet.

There was *the* group of old men – practising their art on pavement slabs under the scorching heat. The weathered tarmac tiles transformed into a grid for al fresco calligraphy. A metal stick that resembled an umbrella rod, with a sculpted sponge in the shape of an inverted teardrop attached at one end, was the 'brush'. Water in a communal plastic bucket became the 'ink'.

Their handmade 'brushes' danced on the ground, making signatures of wisdom and harmony. Their feet, in Chinese black cotton shoes, moved rhythmically with the strokes. Square by square, beautiful phrases flowed out. Stroke by stroke, elegant phrases faded under the summer sun. There were poetic quotes by famous Tang dynasty poets Li Bai and Du Fu; impromptu phrases – in the Shakespearean equivalent of Chinese – that arose in the spontaneity of the calligrapher's mind.

This is calligraphy – a 4,000-year-old fragment of an ancient yet modernising culture. Calligraphy has evolved from pictographs scratched on bones, to a ritualistic art form embedded in China's collective imagination. Each stroke is a fine timeless thread that links – and belongs to – the past, present, and future.

As a welcoming breeze caressed the willows, one of the elders wrote: 'In the carefree spring wind, May the heart be free.' He paused, 'inked' his bush, meditatively pondered over his writing, and wrote: 'When things become most difficult, have courage.'

He looked at me and smiled kindly. 'You understand Chinese, and what I wrote?'

I nodded.

'Would you like to try?'

He lifted his brush.

'Come.'

I tiptoed around the islands of words that separated his feet from mine. I tried to hide my excitement as he handed me the brush.

'What do I write?'

'Whatever comes to mind. Write your name.'

I dipped the 'brush' into the 'ink-well'. The pale yellow sponge absorbed the water hastily and turned a hue darker. *Where should I start?* I hesitated – but my over-saturated brush did not. Water started dripping from its tip, leaving a trail that destroyed the peace of surrounding words. I jumped on to the nearest tarmac square. With brute force, and an awkward stance, I made my first stroke.

The sponge disobediently squeezed out all the water, and created an ink blob the size of three tennis balls. Halfway through my second stroke, I started scratching the ground with my dried up 'brush'. There was no eraser, no backspace. All mistakes, excitement and hesitation showed. Only time would erase my words. Translating my handwriting skills in pen and paper to this tarmac 'canvas', I wrote my Chinese name and a short phrase in an uncategorised calligraphy style.

'Good name, good phrase,' he nodded, 'but too fast. Let me teach you. First, I show you an important word.'

My teacher started writing the character 'Blessing' in the expressive Grass Script, and slowly led me into the world at my feet.

This fascinating story was a finalist in the Bradt Travel-writing Competition in 2010. We have since lost contact with the author.

FAR FROM TIME

Catriona Rainsford

Unpublished Writer Award 2010

'The World at My Feet'

Far away, following the jolting caravans of painted trucks to the south, you reached another world. He had seen it with his own eyes, though not for several years now and the contours of the memory were already starting to blur. But he remembered its wooded hills and wide, sluggish rivers, its tangled streets and chaos of traffic. Most of all he remembered its time. Down there nothing was certain, and everything had to be done quickly in case tomorrow never came. There were clocks whose scowling faces tutted reproachfully at every wasted second, dates that had to be kept, and people who walked fast with their heads bent forward as if chasing the days that time had stolen.

'Time is movement,' I said. 'If every particle in the universe stopped moving, time would not exist.'

He smiled.

In Ladakh there were only two times. Open, and closed.

In open season the warm breath of summer melted the snow from the high passes, clearing the roads through the empty spaces in the roof of the world to the villages and monasteries beyond.

In closed season the frozen jaws of the Himalayas clenched down on Ladakh, their icy teeth severing the one road out of the mountains and leaving the whole region marooned in the sky.

Now it was open, and I moved between seasonal camps where truck drivers and tourists could warm their stomachs and soothe their rattled bodies with chai tea, lentil curry, and salty omelettes wrapped in chapatis. This one was just a few large white tents, round with pointed roofs like circus tops, staffed by velvet-tempered Ladakhis like the man who spoke of time.

I asked him about closed season. When no-one came up, and no-one went down.

'Isn't it frightening? To be trapped all winter in a world so small?' A look of confusion crossed his face.

'Ladakh is not small,' he said. 'Down there it is small.'

He turned to stir a pot which breathed clouds of cinnamon steam into the air, and did not try to explain.

The night draped soft and heavy across the camp, sprinkled with the tiny noises that emphasise quiet rather than break it. The low purr of a generator, the clink of a spoon against metal. Murmured voices, words too hushed to be distinct. The tent was lit by one bare electric bulb, struggling bravely against the weight of the emptiness outside. On benches around the side sat thin-shouldered truck drivers, hunched protectively over their food and stuffing it under their moustaches with calloused fingers polished in grease. They smoked herbal cigarettes from which undulating spirals of smoke uncoiled themselves to mingle with the smell of spices. They mostly fixed their eyes downwards, but every now and again they would look up at the entrance to the tent, where beyond the circle of light cast by the opening the blackness of the night was absolute. Then they would shiver, and draw their jackets tighter around their shoulders with a defensive air, as if trying to keep out more than the cold.

I passed the night in an adjoining tent, veiled from the first by a heavy curtain. Empty of furniture, it contained nothing but carpets. The carpeted floors rose seamlessly into the carpeted walls in a warm cave of geometric patterns. I slept deeply, cradled in the dark richness of their colours.

In the morning, I climbed. I chose the highest of the peaks surrounding the camp and scrambled upwards until the rock fell away on the other side, stripping the view bare to the horizon and leaving nothing between me and the sky.

The mountains rolled away from me in motionless waves of light and shadow. Their sand-coloured backs burnt ochre in the sun and rose to glint gold at their frozen crests, before diving into purple in the troughs of the valleys. They stretched on in every direction like ripples on the surface of the ocean, continually diminishing until the furthest were no more than indistinguishable crinkles on a steely blue horizon and those beyond were swallowed by the curve of the earth. Overhead, clouds loomed like Zeppelins, huge and uncomfortably close, in a sky so saturated in blue it seemed to sag under the weight of its own colour.

Nothing moved. Nothing breathed. The silence reverberated off every timeless rock and sung itself back with the intensity of music.

I stared until my head spun with the vastness of it, and understood.

Ladakh is not small. Down there it is small.

After winning the Unpublished Writer category, **Catriona Rainsford** *went on to write a book for Bradt,* The Urban Circus: Travels with Mexico's Malabaristas, *about her experiences travelling with Mexico's itinerant street circus community. More recently she has worked in human rights in Guatemala, and currently works as a researcher on organised crime. She splits her time between Europe and Latin America and is a keen rock-climber and circus artist.*

JUST VISITING

Chris Baker

Finalist 2020 'And That's When It Happened'

It is not an easy city to love, people told me. Though they didn't quite phrase it like that.

'There's no way in hell I'd spend one minute in that shithole,' were the exact words of one American.

And I could see his point. It is not a beautiful city: a brooding mass of generic concrete, glass and steel flanked by endless rows of shabby red-brick rowhouses and scarred by the giant highways that whisk suburban Americans away from urban decay. And that is not, to be frank, the worst of it. The true darkness that has stained Baltimore's reputation permeates the city in a powerful cocktail of poverty and social exclusion, feeding a never-ending cycle of drugs and violence. With almost one killing per day, on average, Baltimore records more than twice as many murders per year as London. With a population that is less than one tenth of the British capital's. It is a city where you will see small children playing in the rubble of demolished houses; where whole streets have been abandoned to rot and decay; where businesses as innocuous as steamed crab shops place their cashiers behind metal grills and bullet-proof glass, and where, in 2009, the mayor was convicted of embezzling gift cards intended for the city's poorest residents.

Yes, objectively, it is hard to argue with the man who called Baltimore a shithole. It is a city that has suffered; a working-class town that lost its work, where the heavy industry moved out and the grit and grime stayed behind. And yet, when I moved here, I found a city that was somehow unbroken. A city that did not attempt to hide or downplay its problems, but a proud and resilient city that was unashamed of its past and prepared to fight for its future. And to do so on its own terms.

A city where the giant wheels installed to remove rubbish from the rivers draining into the harbour have been given googly eyes and affectionate names; where non-profit organisations have occupied the hulking monuments to industrial America that dominate the most desperately deprived neighbourhoods; where the most celebrated resident is a man who happened to die here while en route from Virginia to New York, and where the second most celebrated is a one-eyed matchstick man from a 1930s beer commercial.

There are pockets of beauty too, among the industrial debris and concrete boxes. From the endless book-lined balconies of the Peabody Library in stately Mount Vernon, to the cobbled waterside alleys that provide paths to America past. From the murals that transform intimidating empty buildings into welcoming works of art, to the repeating rhythms of rowhouses and the salt-splashed quays, this city had got under my skin. I could see its problems, but I could see hope, I could see determination, I could see charm. And most importantly, I could see my place in it all. I had become part of this city, and this city had become part of me.

And then it happened.

I left. Pulled, as if by gravity, to another world. A world that I had also loved, but a world that seemed now somehow lacking. A world that seemed uptight, that lacked vibrancy, that was, well, grey. It took me years to realise that there was nothing actually wrong with this old world; that the problem was with me. That I was missing something.

I can still visit Baltimore, of course, and I do. I am here now, sitting on a bench overlooking the harbour. It is a scene that I have seen a hundred times before; that I can picture with my eyes closed. The sunlight sparkling on the water beneath the downtown skyscrapers; the little boats scuttling between the waterfront neighbourhoods; the stars and stripes fluttering overhead. Not much has changed, it seems, but I know that everything has changed. I may still be the same person, and this may still be the same city, but we are both older now. We have both grown and been transformed, subtly but unmistakably, in the time that we have spent apart. Our lives, once so intimately and absolutely entwined, have diverged.

A woman sits down on the bench, making small talk in that way that friendly Baltimoreans do.

'Are you visiting from overseas?' She eventually asks.

I pause, a wave of memories washing through my mind: of people and places; of friends and food; of laughter and loss. Of a life I loved and left behind.

'This is my city,' is what I want to tell her. But I know that that's not true. Not anymore.

'Yes,' I say. The only thing I can say. 'Just visiting.'

***Chris Baker** is a professional scientist, insatiable traveller and committed islomaniac. Since completing a PhD in computational chemistry, he has lived, worked and studied in China, the United States and the UK. Wherever he has been, whether in the lab or on the road, he has always aimed to build a better understanding of the world around him, and to share that understanding with others.*

THE JOY OF RAIN

Debbie Parrott
Unpublished Writer Award 2015 'Serendipity'

Lightning cracked the sky, thunder followed like slaps of senseless anger.

Elephants; huge, sodden 'boulders', huddled under umbrella acacia as muddy water abseiled their flanks.

We sat and listened to the fat bruising rain…

'Are we stuck?'

In reply, Moses revved the engine. The jeep belched as it regurgitated squelching, gloopy mud. We slithered out of the hole and moved forward.

'Twende, twende: let's go,' breathed Moses.

I softened my grip on the handrail and slowly exhaled.

'Where now?'

'We will go where the track allows us.'

He grasped the wheel as if daring the vehicle to disobey him; he a Masai elder. The rain eased and the stuffy smell of dry earth, swallowing water, floated in the air. The land had craved this inebriation, the drought had been endless. This rain would be welcomed and celebrated in many towns and villages – but not by me. We had set out from the lodge that morning to find a cheetah and her cubs and now my disappointment was dripping from the trees. Nothing for me to celebrate.

The scrub thinned as we neared the plains and, as we drove on to them, the rain stopped and the steaming savannah stretched before us. Like a torch, the sun was already beginning to spread a blanket of light.

'We will have to go to Imbirikani.'

No, I shouted to myself.

'Well… if you are sure?'

'It is our only option until the roads dry.'

The day couldn't get worse.

'We will go and see the village school.'

Obviously, it could.

We splashed down the ramshackle 'High Street' lined with an assortment of corrugated, mud buildings; built from leftovers and strung together like a child's homemade necklace. 'Hotel' was daubed on the walls of one and a doomed goat was tethered at the butcher's shack.

Leaving the village, the road unzipped fields of tomatoes, their newly washed leaves studded with fruit, glistening like red marbles. We arrived at the school; I positioned a smile.

The children were outside under the Kenyan flag singing about a mango tree; apparently, I was lucky because it was not Thursday, Swahili-speaking day. I was welcomed as if I was a rare and possibly threatening animal. We sized each other up. They were a collection of all things ill-fitted: jumpers, woolly hats, pyjama trousers, shoes made of old tyres, odd shoes, no shoes at all. The only uniform thing about them was the mud; they were all liberally splattered. The rain had been thoroughly and messily celebrated and, collectively, they resembled a piece of living impressionism. A large, yellow dress, with a tiny girl inside, sidled up and stretched a fingertip to touch the back of my hand; shocked, she flinched and ran away.

'Come, come! You must see our classrooms, you must speak to them.' Solomon, the head teacher, looked at me hopefully.

Me! Speak? What to say?

He led me into a classroom and thrust a piece of chalk at me; I held it with something akin to horror. I looked around, sunlight struggled through a small window trying to dry puddles left by the rain. The children sat at desks shaped like window frames. A girl, with eyes downcast, showed me her book, every millimetre of each page was filled. A faded, blow-up globe dangled from a nail in the ceiling and the nature-table was growing in the earth floor. The rain would have clattered a rhythm on the roof, I wondered if they had stopped to tap along with it.

Solomon was rubbing a clean space on the blackboard when he stopped and turned to the door,

'Oh no! Oh no! Excuse me. Stay there, I will be back quickly-quickly,' and he ran out.

The boy, with a sleeve unravelled to the elbow, stood up to speak.

'He has gone to chase the animals away, we plant trees and maize but they keep eating them. We need a fence.'

'Does he often run out like that?'

'All the teachers do.'

We waited for him to return. I stood before them, absorbing heat from their gaze; there was a frenzied insect somewhere, also seeking escape.

What on earth can I say?

What can I say… ?

What to say… ?

What to… ?

'Did you enjoy the rain?'

In that dark classroom, dark faces split; it was like streets of glowing windows at night. The children launched smiles that ricocheted off the walls. Balloons of laughter popped and exploded like happy fireworks. The joy of rain fizzed around the classroom. Virtual rainbows everywhere. I wanted to scoop them up; put them in a sealed bag; take them home; keep them; treasure them.

Joy, wrapped up in rainbows, so unexpected; something to celebrate after all.

Debbie Parrott *lives in Guernsey and has travelled widely from early childhood. It began with family camping trips through France, Spain and Portugal to her latest adventure in Papua New Guinea. Only recently has she found time to write articles based on her travel diaries. Four have been published: two in Bradt travel anthologies, one in the* Independent on Sunday *and one in* Writing Magazine.

AN INVITATION TO TEA

Jacki Harris

Finalist 2005 'If Only I'd Known'

I recently visited Mauritania, an Islamic republic in northern Africa. While I waited to be fitted for melt-resistant clothes suitable for the intense summer heat, a man entered the shop and introduced himself. His manners were impeccable, his English passable, and his wife, the proprietress of the tailor's shop. We soon discovered he was the father of four sons, and I the mother of four daughters. After the fitting, his wife agreed to deliver the clothes to my home, as was customary. He offered to provide insight into the Islamic culture in exchange for the opportunity to converse in English and so we agreed to meet weekly.

During one of his wife's subsequent visits I learned she was the second of two wives. He, on the other hand, didn't mention a second wife for several months. Eventually, African hospitality dictated that I be invited to share a meal and meet his family. Despite our burgeoning friendship, the invitation was fraught with misgivings and anxiety on my part, but impossible to refuse.

Might my Western thoughts on polygamy slip past my smiling, but clenched, teeth and spill into our tea-time conversation? What were appropriate topics of conversation with two wives of the same man? Was there cultural nuance, of which I was unaware, which might imply I was interested in being Wife #3? Could I eat food that was unidentifiable? Would I eat food that was identifiable, but not disinfected? I graciously accepted, decided on a gift of two kilos each of apples, bananas, and oranges, and waited hesitantly to be collected.

Wife #1 was beautiful and charming. Wife #2 was the same. My host had excellent taste in women – both friendly and kind. Since I knew Wife #2, she was chosen to cook for the day, while my host and Wife #1 entertained. The ligaments of my Western mind were stretched by this – I doubt my culinary abilities would function under such pressure.

Both wives mothered each of the nine children. Wife #2 nursed the youngest baby, but then Wife #1 cuddled him while we ate. Several of the children were close in age – two boys only one month apart. This African man was either earless, or crazy, to father two newborns and face the possibility of multiple wives experiencing postpartum depression simultaneously. The older children were cheerful, polite, and able to converse, haltingly, in English.

Lunch was served on the floor, following an elaborate hand-washing ceremony including all participants. A large plate of salad was followed by a huge bowl of rice, fish, and accented by small dark meatballs around the platter edge. This was followed by a smaller tray of chicken and French fries. The bowl of rice was approximately two feet wide and eighteen inches deep. Everyone ate from this one bowl with the cupped fingers of their right hands, striving in vain to educate me. Rice and grease was scooped into the fingers, the excess grease drained, and the rice kneaded until it held its shape – then tossed into one's mouth. It was disconcerting, but customary, to have the host and hostesses pick through the meat with their fingers and move the choicest morsels to my side of the bowl.

Along with two wives and nine children, we were joined at lunch by six or seven goats. They were obviously allowed free rein in the family home before entering it one final time as the entrée. There were also two baby goats. European or American children might regard them as pets, but African children celebrate them because of the promise of a choice meal!

Lunch was followed by mint tea, scalding and sweet. I had been warned not to initiate a goodbye until after the third cup of tea. I was not told that my host would delay each cup of tea as long as possible in order to prolong my stay. It was almost three hours before the final cup of tea was served and savoured.

If only I'd known hospitality is not hampered by a plurality of wives…

If only I'd known that children and goats can be proper guests…

If only I'd known utensils aren't obligatory…

If only I'd known not to eat small savoury meatballs – they were made of fish guts…

If only I'd known…

This interesting exploration of culture was a finalist in the Bradt Travel-writing Competition in 2005. We have since lost contact with the author.

RED SHOES

Gill Sutherland

Longlisted 2015 'Serendipity'

Miss Kelsey is the cowgirl I would like to be.

She's tall and athletic, with rumpled sun-bleached hair and desert-wind tan. Like her fellow female wranglers swarming around the corral, matching tourists to horses and giving lardy arses a hefty heave into creaking saddles, she wears a sort of uniform: Cuban-heeled boots, denim, western hat, all lent a matte veneer by the rust-coloured dust that scuffs up around us in the dry heat of a scorching Arizona afternoon in late spring.

Miss Kelsey looks like the love interest from a 1970s western rom com starring Burt Reynolds. A steely-eyed, wisecracking, sharp-shooting kind of girl who's more than a match for mustachioed Burt's deluded machismo. She has no first name that I am allowed to know. Too casual; all the wranglers are known by title and surname only. Respect due.

Despite my aspirations to be like her, Miss Kelsey and I have nothing in common.

This yearning for cowgirl-ness has its roots in my Scottish upbringing. Scotland loves a country ballad, and the sounds of Johnny Cash and Dolly Parton filled my baby ears and beyond.

So here I am on a dude ranch outside Tucson with the other wannabes trying to fathom another identity instead of surrendering to midlife malaise. Mother, wife, daughter and office drudge must make way for inner cowgirl. Well that's the theory.

Miss Kelsey looks me up and down. An appraisement that ends with an abrupt: 'Wally.'

For the briefest moment Miss Kelsey seems to have rumbled my lack of true grit, but then gestures to Miss Smith to bring forth Wally, my noble steed for the afternoon trail ride.

She also signals for the wooden crate that will give my hobbit-like stature a hope of mounting Wally. Once in the saddle, my dignity regained, Miss Kelsey sets about adjusting reins and stirrups to my size. The leather straps slither fiercely like flattened snakes through her gloved hands. I ask her why all the wranglers are women.

'Men wranglers were too grumpy with the visitors,' she replies, herself sounding about as jocular as Clint Eastwood's Man with No Name, but with a twinkle, a wink… then a double-take. She has spotted the footwear I put on in a blind panic when I couldn't find my trainers.

'Is that what you're wearing? Ya got nuthin with a heel?' she asks incredulously, gesturing to my bright-red suede Minnetonka moccasins with tassels and beads I bought yesterday from a shiny mall in downtown Tucson.

'They're quite comfy,' I offer meekly, fearing I've already goofed the first rule of cowgirl club: appropriate footwear.

'OK, Miss Moccasins. You're behind me.'

Our trail takes us through the Saguaro National Park, the cactus conservation area on the eastern edge of the Sonoran Desert. The saguaro cacti are the hugely tall, multi-armed tree-like ones that are used as a pictorial shorthand for all things western. We pass hundreds of them, all with their own identities: some are twisted and old, wizened but wise looking, others proud and fierce, then there's the droopy and comically phallic brigade. Mobs of spring desert flowers,

Mexican golden poppies and the delightfully named fairy duster, blister unexpectedly on scrubby bushes, assaulting the eyeballs with radiant reds, yellows and pinks. Lizards skitter and we watch for (but don't see) the rattlesnakes Miss Kelsey assures us are lurking but probably won't appear until the shadows lengthen.

We pass Paul McCartney's ranch where Linda chose to live out the final moments of her life. Two days before her death she was riding this very desert on her beloved horse. Apparently Paul's last words to her were: 'You're up on your beautiful Appaloosa stallion. It's a fine spring day. The bluebells are all out, and the sky is clear-blue.' After a reflective day on horseback, I sit on the porch of the ranch bar, ice-cold margarita for company, and think about the fragility of life and the hardness of being at one with who you are.

The tranquil desert seems at peace though, stretching for miles in front of me. The night air is still, warm and anointed by wild sage, crickets chirrup and the stars sparkle like rhinestones in the cloudless blue-black sky.

I look down at my sand-blighted moccasins, contemplating their sorry state… And then it dawns on me: it wasn't a mistake to wear them, but serendipitous. Like Dorothy heading home to Kansas, my ruby slippers transport me back to when country rocked our family record player and I wore Clarks red T-bar sandals through the summer. A lifetime's worth of flamboyant and largely unpractical footwear tumbles past my mind's eye: from youthful fire-red Chelsea boots to big-girl scarlet heels. There's no escaping who I am or where I'm from: the shoes give me away. And that's OK.

When not losing the plot in an Arizona desert, ***Gill Sutherland*** *is Arts Editor for the weekly newspaper* Stratford-upon-Avon Herald.

And, yes, she is a Shakespeare anorak. Previously she worked at music titles NME *and* Smash Hits, *and teen bible* Just Seventeen. *She's also written for* Marie Claire *and* The Guardian. *Her three children are especially mortified when she includes them in her ramblings.*

NOW YOU SEE ME

Cheryl Parry

Longlisted 2014 'Meeting the Challenge'

Invisibility, or rather the fear of it, had made me do it. Made me come to Ethiopia for an adventure before I faded away beneath the mantle of middle age. But as I stand at the top of Churchill Avenue, invisibility doesn't seem such a bad thing.

'You. You. You.'

The voices are as insistent as the small hands held out in expectation.

'Mother. Mother. Mother.'

The words as imploring as the brown eyes that follow my every move.

I look about me, panicking, remembering the advice that Roger had given me as he waved me off that morning.

'Walk on. With purpose.'

Easy for him to say – six foot seven, every inch the gentleman traveller, unruffled in his linen suit, a cool head beneath his panama hat – looking imperiously down on the beggars. Harder for me – four foot eleven, crumpled and wilting in the heat and dust of Addis Ababa – trying to avoid eye contact with the children who have stopped kicking a plastic bottle back and forth over the uneven pavement and are now surrounding me.

The knot in my gut tightens; a colonic wringing of hands that has been with me since the journey from Bole airport in a dilapidated blue and white Lada taxi. For three days I have been in a fog of disorientation as Roger conducted a whistle-stop tour. My senses are overloaded by the sights, sounds and smells of this city and its contrasts: the house-of-cards shanty towns bedecked with washing-lines and discarded tyres; the marbled foyer of the Sheraton Hotel with its chandeliers and fountains; internet cafés where teenagers tweet and send friend requests while outside barefooted children play with a Swingball made from a bundle of rags tied to a lamppost; ancient hand-washing and coffee-making ceremonies; the cacophony of the busy roads, the serenity of the cathedrals.

Tomorrow I am travelling south, alone, to a health centre in Yirga Alem to teach ultrasound to midwives. I have no idea what to expect but I realise that in order to be of any use I must get a grip. I have to do this. I have to walk down this street. On my own.

The sun beats down. The brightness of pink and orange coffins stacked in an open-fronted ramshackle shop hurts my eyes. Further down the hill corrugated metal shacks give way to multi-storey buildings trembling in the haze. Horns beep, brakes screech, vendors shout their wares.

The children are still following me.

'Mother. Mother. Birr. Birr. You. You. You.'

Pied Pipering down the road, I decide to take my life in my hands and cross the street at a zebra crossing that is only there for decoration. As I reach the kerb a bus pulls in, its doors gasping as they disgorge passengers on to the pavement in front of me. The conductor smiles as he sings out the next destination.

And still the children clamour.

I speed up a bit. So do they, trotting alongside and in front of me. I'm afraid I will trip as I try to avoid them. Enough is enough.

'Hid. Hid,' I shout, whirling around and around, hoping I have used the correct word to tell them to go.

They stop abruptly, shrug their shoulders and meander back up the hill.

I walk on. With purpose.

I call into the little gallery where Roger had previously bought some wooden carvings. Mohammed the shopkeeper welcomes me like an old friend, asks after 'Dr Roger' and insists that I join him for a macchiato. Like most of the people I have met here he is gracious and hospitable, willing to share what little he has.

There is no sign of any beggars when I go back out on to Churchill Avenue. I walk until I reach a bar. At last, sitting on the terrace with a bottle of cold Castel, I begin to relax. Yellow-billed kites circle high above me, climbing and swooping in a gracefully choreographed display as they scavenge for food. In this small oasis I take stock. It may have only been a walk down the street but I did it and I no longer feel intimidated by the incongruities of this city, of the squalor and splendour, the poverty and wealth, the ancient and modern. I am exhilarated. I can do this. I am visible and I can face whatever tomorrow brings.

But first I have to find my way back to the hotel.

Cheryl Parry *left a career in the NHS to become a (very) mature student, gaining a degree in Professional and Creative Writing at the age of 48. She is currently working on a novel while dreaming of sitting at a window overlooking a brooding sea, tapping out pages of prose till the light fades.*

THE WHALE

Jo Forel

Finalist 2012 'A Close Encounter'

He glides quite elegantly around the river bend, that great blubberish Peruvian in his little yellow pants. And though the current looks angry today, it deposits him gently on the only large flat rock available, as though planned by some great choreographer. He may or may not be breathing.

This, the latest in a surrealist series of events since our departure from Lima, comes with its own set of questions.

Is he alive? Where did he come from? And surely, if superficially, why the pants?

It becomes apparent that the answers can be found pickling themselves to pieces at the bottom of a bottle. He is, as the snatches of whippet-fast Spanish tangle in my ears, extremely drunk. And apparently a fan of swimming.

I look down from my vantage point at the frayed edge of the road, as the coachloads of Peruvians begin to shout and run and act. As they clamber into the river and haul this pisco-soaked soul out of danger and on to dry land like some brave but disorganised human crane. As he lies unconscious under a collection of donated and rainbow-hued jackets, from people who have little to give.

And it's funny, I think, because that's the thing with travelling: as soon as you get on the plane, as soon as you buckle up and accept your first glass of flat Coke from the Lego-haired hostess, you open your eyes. Really open them. You take everything in, however dull or extraordinary. And this man, beached on his rock, the most adventurous traveller of us all, can't see any of it.

He isn't awed by the vast open sky shining down at him, a luxury for me as a Londoner used to packing herself into a drizzling pocket of beige cloud.

He can't see those hills, green with cacti, clenching like giant grass-haired knuckles as they punch their way down to the riverbed.

At the time his eyes started to blur and his words became conjoined, he can't have known I was on the bus, a speck under a cheap polyester blanket, watching a new world through smudged windows. The clean tang of ceviche still prickling my tongue.

He has no idea that the road above him disappeared about a minute before we arrived, with just an hour left of the twenty-one-hour judder to Cusco. That we weren't waiting for him to wash up, but watching the forklift painstakingly move the rocks and debating whether to give up on our coach. That the avalanche is the whole reason we're there. And that he's still here.

Worst of all, he's oblivious as the rescuers pull his friend out of the water ten minutes later, the drinking buddy who wasn't lucky enough to land on a rock. Who splutters and gasps and wraps an arm around him in childlike panic before slumping still against his chest. Before the men with sober faces come to take that friend away, a leaden form in the police blanket, hung between them like a funereal hammock.

He doesn't know he didn't make it.

We decide to abandon the coach and its chain-smoking Scots, busy comparing bites and adventures and what they might consider a tan. I haul my lumpen backpack out of the hold and weave through the toddlers wailing in the heat, the sheep that must belong to someone and the drivers made redundant by the rocks. I clamber down, the road crumbling along with me, just trying to get to the other side, flag a car, get to Cusco.

But the river doesn't scare me now. It's him that's the worry, that slab of man who might have given up. I stumble through the weeds, smell the water as it growls past, reach him. I look. He makes a reassuring splutter and a gurgling snore. A sign that he's ready to open his eyes, come out of limbo, meet the world again. As, indeed, am I.

Jo Forel *is a writer and creative director. She has been published by Dazed and Bradt, led campaigns for major brands including Apple, and taken a year-long research trip around the world from Kathmandu to Cuba with her young family. She really needs to renew her passport.*

TURKANA SANDS

Pat Warburton

Highly Commended 2016 'A Brief Encounter'

Did I *have* to book the cheapest tour to one of the hottest places on earth? I blame my Scottish ancestors. And curses upon that charming hustler in Nairobi. 'Every comfort included.'

What was I thinking?! I'd been conned before in Africa. I could have been in that air-conditioned Toyota that passed us, the only vehicle we'd seen today in this remote and rugged Chalbi Desert, on our way to Lake Turkana.

Instead, I'm slumped on the back of an ancient, dented and defeated overland truck, feeling my brain curl like breakfast bacon. Turkana sun is utterly merciless. (How DO the Leakeys ever work here?)

We've just shuddered to a stop on an empty treeless gravel plain, our third puncture in two days. A collective groan erupts. Group spirits are as flat as the tyre. I'm sweatily hoping the guys will change it; I've had enough greasy, broken-nailed women's lib. It's meltingly hot – I've had cooler saunas – and I'm miserably crusty with dust dug into every pore. The most optimistic genie could not have conjured up the magic of the next minutes, nor its impact…

I see a squiggle. I squint. What is this?! A mirage? A fried-brain hallucination? We've seen nothing living, just heat-shattered lava and whirling dervishes of desiccated earth.

A tweed jacket. A trilby. A lean outline with a walking stick. All coming fuzzily into view through blistering waves of heat.

Disbelief fading, I bolt upright. It had to be him! A legend, and to me, a hero. I'd read he spent part of each year in northern Kenya.

Nearly breaking my leg in my rush to leap from the truck, I manage a sprint to the steadily moving figure and gasp out, 'Mr. Thesiger, I'm so sorry to disturb you, but I've read your books and I think they're wonderful!'

Then, Major Sir Wilfred Patrick Thesiger, KBE DSO FRAS FRSL FRGS, stopped, turned slightly, and replied, 'I'm so glad anyone has read them.'

He'd been in the news, returning to Saudi Arabia, visiting companions from his famous trek in the Empty Quarter. I ask about his trip. He said, 'Things have changed.' I babble a few moments about flat tyres and heat. His face showed, shall we say, forbearance… His eyes were a mixture of world-weariness, some curiosity, perhaps a bit of pleasure. But there was doubt, as well. What was he seeing? A pampered 'pansy', travelling in what he considered an 'abomination'? Or someone who valued his parched and isolated world?

A few more comments. He then nodded, shook my hand, and turned back into the desert emptiness. There was no obvious destination.

Astonished, thrilled, I stand there watching, the sun prickling my skin. This austere man, toughest of the tough, who loved and found peace in the driest, hottest, remotest places, strode off into the shimmering horizon. This man who believed the worth of a journey was directly related to its hardship.

Directly related to its hardship… That phrase niggled itself like an annoying fly into my mind, then gnawed into my marrow. Shaming me. Challenging me.

The next days held many joys. Lake Turkana, that spectacularly coloured indigo-jade-emerald-teal splash on an endless palette of tan, enchanted me. There were glimpses of befeathered Rendille, head-scarved Borana, splendidly beaded Samburu, skinny El Molo. Exotic camel herders, Somali ostrich haughtily fluffing their plumes, night skies holding glittering stars inches from my nose, the dark drama of the jumbled and jagged volcanic hills… Extraordinary, all; but always shadowing me was the doubt in those weathered eyes.

The niggle didn't let up, either. Scorching temperatures, rice seasoned with grit, ghastly roads, choking down hot chlorine water. Each a chance to pony up. Each a chance to become something better.

Hot places still crook their come-hither fingers at me; I can't resist. The Afar, the Taklamakan, Ghademes, the Flaming Mountains. Their lonely beauty calms me. The sun broils, sweat streams, dust sticks. I don't blame. I don't complain. I've made my choices. I want to remove the doubt I saw in those eyes.

Even now, twenty-five years later, I sometimes laugh as I shake out my Therm-A-Rest. I know exactly what Thesiger would say! He famously said it to Eric Newby in the Hindu Kush. I endure the disapproval. But I smile a little too, because I think the doubt in his eyes would be gone.

Pennsylvania farm girls don't often get to all the continents, but ***Pat Warburton*** *always knew she would!* National Geographic*'s arrival each month was a highlight of her entire youth. She chose to be a dental hygienist so she could travel, and luckily she loved the profession. After washing ashore on the Falkland Islands a few years ago, she's still practising happily in the capital, Stanley.*

HOPE IN PINK MERINGUE

Anita King

Winner 2021 'I'd Love to Go Back'

I am greedy. I want to go back to Damascus, but I want to go back as it was before the war, before our screens were filled with its anguish, and before the tormented numbness of old men and young women shattered my sleep. Had I passed some of them in the tiled courtyards and arched alleyways in that long-ago time? And what had become of Amira?

I always wanted to go to Syria. I had grown up with tales of the Old Man of the Mountain, of rich libraries and intricate carpets, looted or burned by marauding Crusaders. An uncle with a love of history and theatrics in equal measure had returned from travels to Syria with slides of mountain forts, where, he declared with a flourish, Salah al-Din had triumphed over the heretics. My uncle's tales were recounted over the whirring of a slide projector, accompanied by the occasional brandishing of a carved dagger; all of which left us children with goosebumps and a fair few nightmares.

But in the end, it was not a carefully planned and researched trip with guides and pre-booked hotels, nor a notebook with my uncle's contacts. Rather, the decision to go to Syria that November was a bit random, and we left almost as soon as the visas arrived. We would fly into Damascus and back from Aleppo, and make our way between the two cities by whatever means seemed best. Not our usual way of travelling to intrepid destinations, but we were bereft and depleted, and longed to be overwhelmed by something other than sorrow.

The source of this morbid state of heart and mind? The sweetest baby in all the world, with a dainty mole beside her left nostril, and soft brown hair flecked with gold. She had been our daughter briefly, and then she was no more. In our grief, we doubted we could ever again put ourselves through anything that could leave us so utterly hollowed out.

The sixteenth of November 2007 is the date I would choose, if I could return. It was our third day in Damascus. We had spent the afternoon wandering through the Al-Hamidiyah Souq inside the old walled city, after eavesdropping briefly on a guide at the imposing Temple of Jupiter. In that moment, though, I could not have been less interested in any pile of stones, Roman or otherwise.

The souk, however, drew me in. Slowly, I became immersed in the colours of glass lanterns and sugar-coated sweets, and in air thick with the aroma of perfume, green herbs and ground spices. It was Friday, and there was a celebratory vibe, an exuberant chattering cacophony. Young women in short skirts, with arms linked, sashayed between older women in embroidered abayas and black gloves, while toddlers darted between their fathers' legs and mothers' skirts.

Our destination was Bakdash – one of the few recommendations we had jotted down before our hasty departure. 'The very best ice cream in the Arab world,' a Syrian friend in London had insisted, her voice infused with nostalgia. 'They pound it by hand, and roll it in layers of the freshest pistachio.' So of course, we said we would go.

And there, framed by the wooden doorway of Bakdash, was Amira, in a candyfloss-coloured pink dress like a gigantic, cascading, meringue. Pearl-like beads were scattered over the already overloaded hemline. White patent shoes with pale pink butterflies, and white glittery tights, completed the confection. A large silver badge pinned to her front, with the number 5, announced the occasion. When I saw her, she had just caught sight of herself in the gleaming glass doors and pirouetted in delight, her brown hair with gold flecks dancing in ringlets around the most kissable dimples.

I stood. Transfixed, remembering… and imagining. Amira threw her head back and, unexpectedly, our eyes met. To my own surprise, I reached out my hand. With a giggle, she reached back. Her fingers were sticky. Maybe that dress really was made of candyfloss…

I felt tears starting to fill my eyes. I blinked them back, but she had noticed and looked suddenly bewildered. Before I could speak, though, a voice called out her name. As I watched, she turned to join the gathering birthday guests. She did not look back.

Amira will now be 18 years old. My daughter, born a year later, is 12. Her hair is darker than Amira's; and even at age 5, she would never have countenanced pink lace. Sometimes, though, when she smiles, there are faint dimples. In those moments, I find myself thinking of Amira, wondering where she is, and, always, sending her silent thanks for the hope she so unexpectedly offered me that day.

Living in different countries from an early age as her family moved around has given ***Anita King*** *an appreciation for 'slow' travel, and the pleasure of everyday encounters. Over the years, she has spent extended periods of time in places as diverse as France, Canada, Kenya, the UAE, Tajikistan and Pakistan, all imbuing her, she feels, with a western head and an eastern heart! Travel writing is new for her, but she is definitely smitten!*

THE DREAM OF THE DESERT

Agnieszka Herman

Highly Commended 2021 'I'd Love to Go Back'

The rock shelter is very small. It has a wide, lens-shaped entrance and an uneven bottom sloping towards one of the side walls. Outside, bizarrely shaped towers of weathered lava loom in the flickering air, dark and sullen under the merciless, white-hot desert sky. No life. No sounds. Inside the heat is less cruel, but we have not come here for the shade. We've come for the antelopes, giraffes, for men with bows in their raised arms and women holding hands, for camels, cattle, for hard-to-identify objects used for long-forgotten activities, for overlapping, interpenetrating signs and figures covering the walls, so numerous that many of them will become visible only much later, at home, in enhanced photographs processed with specialised software. Some of the images are sophisticated, ingenious, other schematic, misshapen, some are faded, others so sharp that they seem fresh. The most mysterious ones bring to mind wide-eyed ghosts with outstretched arms and boats with strangely shaped sails. In this surreal, dead landscape devoid of humidity the association with boats seems absurd. It's what our modern European minds see, the minds of strangers. And yet, all those animals, women and men, a long time ago, here, exactly at this spot – they must have had water.

Long time ago.

Time.

A very illusory concept in a forgotten cave at the foot of the Emi Koussi volcano, in the middle of the Sahara Desert. And it is not just a subjective perception of someone who's lost their sense of time while trying to reach this place. Of someone who has spent a week in a hot car incessantly getting stuck in the sand, who's experienced an overdose of the stunning diversity of monotonous desert landscapes, countless rows of elegantly shaped barchans, rolling plains of perfectly sorted stones, impossible human shelters braving the sand in the middle of endless, flat expanses of nothingness. On the way to Tibesti time disintegrates gradually among the dunes of treacherous ergs, along stony *pistas* winding through minefields and between rugged rocks dotting the bottoms of paleo-lakes.

At the back of the cave there's a small opening in the ceiling. A beam of light falls on a twine net bag stuffed with rags, hanging from a stick plugged into a fissure in the wall. Who left it here and when? Impossible to tell. In the hyper-arid desert environment things discarded or forgotten yesterday and years ago look exactly the same. Potsherds, bones, cartridge cases, wrecked tanks and cars, stone tools, they all lie covered with sand or exposed to the burning sun, indifferent to the passage of time. Eternal.

The night before, after a supper of pasta with nauseatingly smelling meat you wouldn't dare to eat under other circumstances, we lie on our backs and watch the rocky towers surrounding our camp cut out black geometric shapes from the starry sky over our heads. Mahdi, our guide and guardian, tells us the story of this place. A story full of trees, villages teeming with life, animals hiding in the thicket.

'Right over there, see?'

'When was this, Mahdi?'

'I told you. When my father was a child.'

'And how old are you now?'

'56.'

It might be just a random number, a careless answer to one of those annoying questions we keep asking him: how many? when? how much? Mahdi might be older. But still, is it possible that all those things have disappeared, that the whole landscape has transformed so drastically within just two generations? No, not even here, in the part of the Sahara where climate change is occurring faster than elsewhere in the world. But then – is this story a metaphor? Does 'my father' mean 'my ancestors'? Or is it all just fiction?

Now, stuck at home during the pandemic, I'm reading that the rock closest to our camp is called Koubou Dougouli. Inhabited place. Site with water. On the map I have it is the only peak in the whole area which has a name and an altitude, although it doesn't differ from countless similar peaks around it. Does the name reveal the history of this place?

If the relatively recent past is so inscrutable, so elusive, how can we say anything about the time when the rock paintings were created? They cannot be dated, all we have is presumptive evidence. And stories. Stories that hook you with their simplicity. In the minimalistic world of the desert small details grow and demand your attention with an intensity you don't experience anywhere else. They remain etched in your memory like engravings on a rock face. You don't leave this world unchanged. You want to come back.

I bet the wooden necklace I hid in a fissure in the painted cave still looks the same when I'm there next time.

Agnieszka Herman *is a physical oceanographer studying sea ice and polar oceans. When she isn't busy breaking virtual ice floes with computer-*

generated waves or spying on sea ice in satellite images, she goes on real (and preferably long) trips to remote places. Mountains and deserts are her favourite habitats. She lives in Gdynia, Poland.

7
GHOSTS & DUST

"Here at my feet is a little patch of the world, which connects directly to my past."
Marie Kreft

Sometimes we are looking for connections. Perhaps we travel to find ourselves or our roots or our families? Sometimes we stumble upon history – or into it. All of these stories have links with the past – they stir up ghosts and dust.

Lithuania Poland Spain Japan UAE Chile Italy Greece UK

CROSSES

Sara Evans

Winner 2005 'If Only I'd Known'

A simple wooden staircase snakes up the green hill in front of me. Countless Lithuanians have been up it before me. From towns and villages, they have travelled through the dense pine forests that typify central Lithuania's evergreen landscape to this small hill surrounded by grass fields. Once here they have planted crucifixes. Thousands of them.

This is the hill of crosses. It's estimated that well over a hundred thousand crucifixes stand on this hillock. Rising up and over the hill, they spill endlessly into the meadows below. Almost vacuum-packed, there's nowhere else on earth where so many crosses can be found all in the one place.

As if in holy gridlock, crosses, large and small, vie for space. Under large crucifixes, smaller ones are stacked high in unsteady piles. Off the arms of crosses hang yet more crucifixes, tiny and fragile, wedged tightly against each other like clothes on a sale rail. And, where a small tree or bush has managed to grow, crosses are draped off these too or planted under their boughs.

Criss-crossing through the crucifix tangle to the top of the hill, the variety of crosses is amazing. Mass-produced, cheap plastic crosses are heaped next to those handcrafted from oak, painstakingly engraved with scenes from local folklore. Impromptu crucifixes made from matchsticks, held together by chewing gum, lean against large mock gold crosses that glitter brightly with fake rubies and emeralds.

It's as if a hatful of crucifixes has been dropped from the sky, there's no pattern to the way the crosses have been planted. Small or tall, cheap or expensive, kitsch or sublime, the crosses here are intermingled, jumble-sale style.

Reaching the top of the hill, the wind picks up and an eerie sort of music starts spinning around me. It's the sound of thousands of rosary chains, hanging from crucifixes, singing like wind chimes. There must be millions of beads here; sprinkled all over the hill in rainbow colours like 'hundreds and thousands' on a cake.

As I make my way back down, the wind drops and it's cemetery-quiet. The statues here make mute company. Wooden angels perch child-like around the hill. Virgin Marys smile in their plastic shrines. Life-sized characters from Lithuanian legend stare forlornly on. Their silence travels with me until I reach the road at the end of the hill where the sound of traffic breaks their spell.

At the bus stop I sit and wait for my bus. After the hill and all its treasures, the fields either side of the road here seem incredibly empty. An old lady arrives. She sits slowly down next to me. A big, black shawl, drawn tightly in, keeps her warm. A brown headscarf holds snow-white hair back from her face.

From a worn-looking bag, she takes out a flask and pours black tea into a little cup. Sipping the tea, she asks in quiet, broken English, if I have been to the hill of crosses. When I say that I have, she wants to know know if I've planted a flag.

'No. I'm not religious,' I reply.

'You don't need to be,' says the old lady, her English growing more confident.

'Do you know of the hill, of its history?' As I shake my head she, like a folktale babushka, tells me the story of the hill of crosses.

I learn that the hill – once a site of pagan worship – has, for many Lithuanians for hundreds of years, been the focus of national defiance.

Crosses were first planted here after bloody uprisings against Tsarist rule. Continuing into recent times, when Soviet purging killed and exiled thousands, planted crosses have represented fallen countrymen and an enduring commitment to a free Lithuanian state.

As my bus comes and leaves without me, I discover that, on at least three occasions, the hill of crosses has been bulldozed to the ground. Iron crucifixes melted down. Wooden ones burned. After each razing though, Lithuanians kept coming back, risking their lives to plant their freedom crosses.

Then, with the fall of the Soviet empire, came independence for Lithuania. The hill of crosses was left alone. Today, crosses are still planted on the hill. People travel not just from the Baltics, but from all over the world to add their cross, their own mark of respect. As another bus arrives, I thank the old lady for sharing her true story with me.

On the hill I had been beguiled by the outpouring of religious eccentricity. But I was just seeing and not understanding. I should have planted my own cross. Acknowledged, like thousands of others, that inside this small hill, lives the spirit of a people as strong as a mountain.

***Sara Evans** is travel writer published by the Saturday and Sunday* Telegraph *newspapers,* The Guardian, The Independent, BBC Wildlife Magazine *and* Africa Geographic, *among others. She is also the author of the critically acclaimed* When the Last Lion Roars *(Bloomsbury). Sara has also been a panellist at the Royal Geographical Society and guest speaker at various UK events and radio programmes around the globe.*

RUMIA – A LOVE STORY

Marie Kreft

Winner 2010 `The World at My Feet'

We're skulking along the shadowy side of a street in Rumia, north Poland. The sun's burning a hole in the back of my neck and Steve is cross that we didn't print out a Google map to guide us to the cemetery. I stop inhaling the sweet warmth from the summer hedgerows in case it later becomes the scent of disappointment. There may be songbirds, but I'm not listening.

'Oh I'm the one who's grumpy?' says Steve. 'No. I just think we should've researched this better before we set off.'

Rumia is the birthplace of my late paternal grandfather; the resting place of three of his brothers and their mother Kunegunda. My fiancé and I have taken a few days' detour on our overland journey through Eastern Europe so I can see where they are buried.

But we have less than two hours before we must catch a train back to the seaside resort of Sopot and I'm chewing my knuckles in worry that we won't find the grave.

'Hold on, what's that?' I spy a church steeple among treetops in the distance and my heart leaps up to nestle in my throat. We hurry towards it, passing simple brick houses, a drip-dropping stream and a statue of Pope Jan Pawel II with his arms outstretched in imitation of the cross.

But the church and orderly cemetery look too modern and my heart falls to its normal place.

'Shall we take a quick look anyway?' I suggest, in the absence of a better idea.

We split; Steve surveys the left-hand section of the cemetery and I take the right. I wander up and down, row by row, studying the names of the occupants of each grave.

Some plots are recently dug, with shiny marble steps, coloured lanterns and lilies still in their cellophane, while others are older with weather-cracked headstones and the occasional weed making a break for daylight. Three women are spending their sunny afternoon tending to the resting places of loved ones, scrubbing dirt from memorial slabs and trimming grass around the kerbs.

My hopes lift when I realise how many Krefts are buried in the cemetery. Kreft is not a common Polish surname, but here lie the remains of scores of us: Wiktor, Aniela and Jan, Marta and Feliks, Hubert, Brunon, Stanislaw and Elizbeta. Even Romauld Kreft, who was born in the summer of 1955 and dead by winter. Maybe some of them are my relatives.

I'm lost in my thoughts, enjoying the sunshine and dreaming up hopelessly sad stories about poor baby Romauld, when I see Steve waving arcs in the air. I skitter up beside him and he whispers, as though not to wake anyone up: 'Is this the one?'

It's her. There at my feet lies Kunegunda, 1884 to 1955, with 15-year-old Lucjan, my father's namesake Leon, and Kazimierz, who was arrested in 1952 and never seen alive again. Their mottled grey headstone has a carved cross at its centre, from which Jesus's face stares mournfully down.

The cross could be depressing to look at, but even the frown of Jesus can't stop me from smiling. Here at my feet is a little patch of the world, which connects directly to my past.

This is the grave of my great-grandmother and great-uncles, and their bones in the soil are my solid link to Poland. And an explanation

– or perhaps an excuse – for my love of pieroghi, the smell of fresh dill and the crisp taste of zubruwka, bison-grass vodka.

Maybe they're why I load my breakfast plate high with pickled gherkins in Polish hotels, and giggle when Steve turns pale at the sight of them 'so early in the day'.

There's a second reason for my smile: someone has recently visited these Krefts. Three pots of silk daffodils guard the slab in front of the grave, their fake petals still unfaded by the sun. 'I know what I want to do.'

Steve guesses straight away. A few metres away from the church is a florist's shop with a lucrative sideline in graveside memorials: candles, lanterns and china cherubs. There I buy a modest pot plant with red blossoms and succulent leaves and carry it back to Kunegunda's resting place, where I set it down next to the daffodils.

'A small mystery for the living Krefts of Rumia,' says Steve, and I'm filled with joy that having found an anchor to my past, I'm journeying onwards with the man who is my future. 'Thank you.' I hug him and he leans forward to whisper into my hair. 'Next time,' he says. 'Next time can we please print out a Google map?'

Since winning the competition in 2010, ***Marie Kreft*** *has written and published two editions of* Slow Travel: Shropshire *for Bradt Guides and several articles in national titles, including cover features for* BBC Countryfile *magazine. Marie runs a copywriting business. She married the Steve who accompanied her to Rumia and they now live in Birmingham with their two Brummie boys.*

GIRONA'S CITY WALLS

Gabi Reigh

Highly Commended 2017 'Lost in Translation'

'Are you sure you still want to go?' my mother asks me, 24 hours before I fly to Girona. My mother worries about everything, but even she would not have considered worrying about me going for a short break to one of Catalonia's most elegant cities until a week ago. Before then, for the tourist, Girona only had two faces: as the 'gateway to Costa Brava', proudly proclaimed by its airport, in letters large enough to see before you hit the runway, and as an intriguing cultural destination for anyone searching for the 'new Barcelona'. It was the second of these Gironas that I was hoping to find, yet a week before we arrived there the city had taken on a new identity as a place of struggle and rebellion following the Catalonian independence referendum. To my mother's concern, Girona had made front page news as thousands marched in the streets protesting against the actions of the Spanish government.

We didn't know what to expect. The news distils the momentary passion and drama of a particular place, and you imagine you will see it there still, days later, that energy imprinted on its streets, on the faces of its people. But Girona, on that first morning, was full of tourists revelling in sunshine and locals chattering in animated Catalan in cafés clustered around stately squares. Yet as we rambled around the medieval old town, in every cramped alleyway we saw defiant 'Si' posters, urgent claims for independence, and, on banners, mouths taped shut, speaking of unresolved arguments, conflicts waiting for a spark to rekindle them again.

The first thing that any dutiful tourist will do in Girona is walk around its city walls. Built by the Romans, then besieged by time, the walls were fortified in the Middle Ages to lock out the threat of the world. We walk around them at first, unable to find a way in, shut out like Moors by those imposing fortifications. Finally, we climb up, and look down on a Girona once safe and protected from its enemies. We walk through its history, behind us the dark labyrinths of Barri Vell, the old town with its moon-white cathedral, past bridges where Modernista houses weave a red and ochre patchwork over riverbanks, towards modern apartment blocks where Catalan flags share the wind with untidy laundry streaming from balconies.

Bridges and walls. Isaac Newton once said that 'we build too many walls, and not enough bridges'. I don't know how to read Girona. I don't know if these flags, these banners, raised up as earnestly as these enduring stones were once glued together, are walls between this city and the rest of Spain, or whether they are bridges between its people, a bond of solidarity against the oppression of the past. I have no-one to ask; I walk outside these lives, a tourist following a trail above the city. What's more, communication is a problem. Diligently, I had been storing crumbs of Spanish from other trips, and could now proudly consider myself competent enough to ask for the bill in a restaurant or extract information about the location of the train station. Yet now, these phrases too, instead of building bridges, raise up walls. People respond in Catalan, language yet another banner of identity, savoured by Girona's people after long years of Francoist suppression.

Beneath the walls, I hear the city beating with its many hearts. We pass playgrounds, full of speed and voices, we smell, before we see, heaving paella pans brought out to impatient restaurant tables, and in a dim courtyard, an old woman, slowly walking. I stop to look at her.

We are alone, my stretch of wall suddenly empty. In her courtyard, the dark tendrils of an overhanging tree stretch out between us. She is moving one step at a time, and I can hear her slippers scraping ancient stones, her walking stick tentatively searching them, like the horns of a snail, guiding her step by step to safety. A distant church bell rings, paling under the scraping sound. She is history, snailing her way through decades of war, dictatorship, into an uncertain future. Unseen, I watch her for a long time, yet cannot see her destination.

***Gabi Reigh** moved to the UK from Romania at the age of 12 and now teaches A-level English. Her short story 'It Was a Very Good Year' was shortlisted for the Tom-Gallon Trust Award in 2018. In 2017, she won the Stephen Spender Prize for poetry translation and this inspired her to translate more Romanian literature as part of her 'Interbellum Series' project.*

YES. NO. IT'S A LITTLE DIFFICULT

Chris Walsh

Finalist 2017 'Lost in Translation'

An example:

At the start, when I first went there, I might have said, without thinking much about it, 'the colour you are looking for is is "red".'

Now? I'd suggest differently. Crimson. Rose. Cardinal. Salmon. Fire truck. Scarlet. The colour of old blood. Fresh blood. Your blood.

Shades and hues. Nuance and subtlety. It's a difficult thing.

No, it's not.

It's… the only thing.

* * *

I liked Yoshie from the scorching July we first bowed to and at each other. She would sit and angle her slight and ageing frame to be sure to catch my movement, gauge my disposition, and try to please the confident educator that I was then masquerading as.

Though Junko, Toshiko, Chie, and the few other students whose names elude me now, also comported themselves in some response to my quirky language practices and wishes, they always held Yoshie, their group's natural leader, in their peripheral view. For if she twitched for better or worse then the whole group spasmed and shifted mood together and I had not much significance in that process.

I had swung into Japan earlier in the year, had been interviewed by the management of a language school of dubious history and uncertain destiny, and offered a shaky role handling an array of erstwhile scholars who were following the trend of trying to pick up someone else's native tongue and culture by interacting with visiting foreigners.

Included in my disparate mélange of Wednesday students were Yoshie and the rest of the 'Golden Agers'. An ancient crew – more motley than masterly – of learners, they had my attention and awe from the start. They were 'hibakusha' – survivors of the atomic bomb. Helping them felt like solemn privilege rather than duty.

Having come through the infamous devastation of their city and decades of post-war stress these women were never ever going to be troubled or shaken by any classroom duress that I could subject them to, I thought. Surely.

Somehow, when we are younger, we get these things wrong. Our timing is awry. An older me wants to now, occasionally, chide a youthful me.

A 'sensei' – teacher – in Japan is accorded a clear deference or obedience, and all kinds of leeway that educators in other pedagogic worlds may never know. So a sensei asks and a sensei gets because they, their age notwithstanding, are 'the ones who have gone before' and are expected to know all sorts of things about measuring twice and cutting once.

In early August, in the days following the A-bomb Memorial Ceremony near the device's Ground Zero point, I asked for, firmly though not unkindly, my Golden Agers' first-hand narratives of what happened on and about the 6th of August, 1945. Ever so slightly they pivoted towards Yoshie. It was her say-so. Her lead.

She lowered her voice, tilted her head to one side. She paused and breathed this: 'Sensei. Chotto muzukashii desu.' Teacher, it's a little difficult.

In the years that followed, I was to learn exactly what this meant. It was a clear but polite 'no', in Japan. But I was not quite ready or able to hear that yet. After all, 'difficult' was not 'impossible', it seemed.

'If you tell me your stories I will feel your experience better.'

'If you share your narrative I will faithfully pass it on to the generations I will meet in my future teaching life.'

'If you help me understand what it was like I may know more about the futility of war. Can teach the futility of war.'

'If you do this thing.'

If YOU do this thing.

And we began.

In the following hours I learned of mayhem, noise, fire, pain, stench, horror, shock, dread. Of loss, uncertainty, pathos, lethargy, futility, grace, fatigue, desperation, luck, generosity. Of the business of being appalled.

Of grief, of grief, of grief.

When the time was up, when tears were wiped and the tissues discarded, they left together. Smiles back in place. Lace parasols and paper fans set to handle the fierce things that the Hiroshima summer can surprise the unwary with if they're not setting sentries for them.

Two weeks later Toshiko succumbed to the radiation-induced leukaemia she'd done battle with all the time I knew her, and many years before that. Nobody in the class said a word about it. They merely tilted her chair against the table edge and we went quietly through our afternoon's practise and never mentioned atomic bombs again.

Just so. It had been a little difficult.

First a school teacher and a language teacher and tutor, then a travel agent, then a bunch of other things that made it possible, ultimately, to travel the world, ***Chris Walsh*** *has taken himself all over the globe. From his native New Zealand, he has lived in nine countries and had numerous adventures in more than sixty, writing and photographing along the way.*

THROUGH THE BLUE

Hannah Doyle

Finalist 2019 'Out of the Blue'

It was back in the old days, when Grandad was alive and Dubai was just a dusty old desert town, back when I moved like a silverfish, darting under rocks and slipping through the sand.

Grandad would take his Pointer roaming through the scrubland that backed on to our house, where rangy, sparsely feathered chickens squabbled in the dirt that passed for a garden, and one-eyed dogs prowled the streets in packs. They have gold markets for tourists there now, glinting skyscrapers and malls. These were the old days though, with nothing to do but swim in the sea and climb the palms, hiding in their splayed fingers and plucking dates on the fly.

One evening, Grandad came home early from his work at the construction company, his fleshy hands two of the many that were busily building a glimmering new metropolis for the modern age.

'Get your shoes on,' he said.

And so I did.

We stepped out under an electric, deep-blue sky, the sun dying in the distance and bleeding rosy through the honey locust trees. Our Pointer cantered ahead, skilfully avoiding the thorns that jutted from the trunks, and I too followed my own scent trail.

'Don't go too far,' Grandad warned.

'I won't!'

The trees here were tightly packed and spiked, their slender leaves floating like lily-pads in the warm air, and me drifting under the surface. The crease behind my knees wept sweat as I played catch-up with my looming shadow, growing so engrossed in the game I lost track of the time.

'Grandad!'

No reply came. I turned on myself, looking this way and that, but it was no use. Grandad was nowhere to be seen, the thickets were too dense, the spines too clustered, to see very far ahead. I paused then, and a creeping sense of worry tiptoed over my shoulders, uncurling over the base of my neck. If night fell, what then? What crawled out of the dunes once the sun had gone? What lived there?

In the distance, a wailing started. The evening call to prayer unfurled over the desert, a haunting cry that seemed to enchant the very stones beneath my feet. The muezzin unleashed his keening, sending it out and up through the air, summoning the trees and sand to attention. As the call soared higher, I glanced up at the sky, watched it pull down a dusty veil that turned the whole world and everything in it blue, a filter that would soon shift violet and inevitably fall dark. I pressed on through the trees, tentatively now, placing one uncertain foot after another, fighting back worry. And as I moved through the blue, the creature appeared from within it.

It stood under the tree on spindly legs, its thick-lashed eyes deep and dark and gentle in the fading light. Although taller than me, its muzzle was tilted, its head arched back as if in assessment. Wrapping its meaty tongue around the locust leaves, it ate slowly and carefully, eyeing me as a child might a new playmate, but unafraid. I had seen camels before. I had ridden them, even, but always bridled and saddled, moored to their owners who dragged them with one hand and clutched their dish-dashes with the other. Alone in this small break among the thorny trees, the camel towered over me, an earthy smell of musk and sweat rising off its coat. As the call to prayer tailed off, the camel ended its feast. It turned to look at me. It seemed to me in that moment that the desert simply evaporated. As I stared deep into its eyes, I felt a breaking in me, a sadness I didn't understand. And then as unexpectedly as it had appeared, the camel left me, gliding away through the trees, hooves sure on the dirt.

For a few minutes, I waited there alone, before turning back the way I came. It wasn't long before the dog found me.

'Grandad!' I ran towards him, thrusting my hand into his.

'I saw a camel,' I said. 'A camel!'

'Did you really?' Grandad slipped me a side smile.

'Yes,' I frowned. Didn't he believe me? 'Really.'

'Don't go off on your own like that again.'

But I did.

Years later, I left Dubai. The city grew up and I did too. I never went back.

It had existed at the frontier, that camel. The frontier between the old and the new, the wild and the tame. It had not been afraid. It had stood at the gateway to a sacred space, the space in which a child first realises the immensity of the world and her smallness within it, a fleeting glimpse of the end of time, waving me across the border and into the blue.

***Hannah Doyle** is a writer and translator who grew up in different countries and now splits her time between the UK and France. Her translation of Marc Levy's* Hope *was published in May 2021. She is represented by Hellie Ogden at Janklow & Nesbit.*

WHAT WAS LEFT BEHIND

Jane Westlake

Longlisted 2013 'A Narrow Escape'

The studded wooden doors weren't a feature in the walking tour of Santiago, but when the guide paused briefly beside them I felt a sharp tug of recognition followed swiftly by the dull ache of memory.

More than two decades before, I'd flown into Santiago seeking some rest and recuperation. I'd been working on an archaeological dig in the Urubamba Valley in Peru; this time in Chile was meant to be a holiday. I'd caught a taxi from the airport and negotiated the fare but let the driver put my rucksack in his boot. Outside the guesthouse I handed over the agreed fare but no, he wouldn't open the boot unless I paid him substantially more. A trick which damned the whole of Chile in my eyes, at least for those first few minutes.

After sleeping on a rasping Lilo in a tent for months the smell of furniture polish which greeted me at the Residential Londres was a great comfort. Sturdy wood enveloped me: on the panelled walls, in the chunky parquet flooring and lace doilies were placed purposefully on dependable oak furniture. I felt secure and relaxed as I peered from the wrought-iron balcony of this stylish 1920s building in Barrio Paris-Londres. Should I continue to carry my passport in my money belt or maybe it would be safer here? I hid it and locked the door.

I sipped an espresso standing at a tall table in a café in the Paseo Ahumada. So far, so continental. The well-groomed businessmen exuded satisfaction and good cologne. It all seemed so familiar: carts in the streets selling wriggly churros like the ones I'd crunched my way through in Salamanca, European-style buildings with a flounce of baroque or neo-classical grandeur. Then one window display hinted at the reality of life in Santiago that day in 1985. Male shop dummies in expensive casual wear were posed with machine guns. A couple of the life-size replicas were pointing at me. What sort of country was this were guns were used to sell clothes? The military coup may have been in 1973 but General Pinochet was still in power.

I was looking for somewhere to buy some wine to take back to my room when I heard the chant. A rambunctious rhythm stamped out in a nearby street, '*El pueblo unido jamas sera vencido*' (The people united will never be defeated). Less than a year earlier I had lost someone close to me and was still in that strange shifting land of bereavement where anger can turn to euphoria and lack of concern for one's own welfare. The vibration of that life-affirming chant drew me in. I found the demonstration; I had no ID.

The fact I was wandering alongside rather than in the centre of the sweaty exuberant mass made little difference when the mood changed. Riot police swung into position from adjacent streets, batons were raised and the release of tear-gas from a small tank was met with shouts of Zorrillo or Skunk from the scattering crowd. I had no hanky or lemon to provide some respite from the smothering, clawing effect of the tear-gas on my lungs. Even with streaming eyes I knew I had to move. I was running beside a bank with studded wooden doors when a hand reached out and I let them pull me in. When do you trust a stranger? I learned later that some demonstrators had been rounded up and taken away, maybe to join the 'disappeared'.

The middle-aged bank clerk who had reeled me in was surprised that she had caught a tourist. I had been the last one in before they bolted the door. A confession of my stupidity in not carrying my ID, particularly when people didn't know where I was, was glossed over with chat about family, hers and mine. A very ordinary conversation in what for me at least was an extraordinary situation.

So my attention had been drawn by these doors, and it was only when the guide referred euphemistically to the 'change of government in 1973' that I was aware of him again. It was as though he was rewriting history. It was pointless me standing there. The doors were ajar so I pushed one and walked back in.

Jane Westlake *used working on archaeological digs abroad as a way of bringing adventure on a shoestring into her life. Inspired by freelance journalists she met while in South America, she went on to work for the BBC as a radio feature maker and as a TV and radio news producer. Jane lives in Brighton.*

PRISONERS OF POMPEII

Sally Stott

Finalist 2015 'Serendipity'

Somehow we had ended up in the cemetery. It was the last thing I wanted to see – as well as the painted house, the sauna, the dog mosaic, the brothel with sex graffiti, and the room full of broken pots. Just a few more things on the map! Only now it was dark, the tombstones, as tall as houses, were looming over our heads, and we couldn't get out.

'Gates close at 7.30pm,' a sign had said in Italian, but the only Italian we knew was 'ciao' (hello) and 'ciao' (goodbye), so we hadn't seen that. As the sky turned to black, Pompeii, a skeletal city immortalised in ash, opened its burnt-out eyes, and we realised something: we were the only people still here.

We ran past the graves, along the bubbling stone of never-ending streets – which a few hours earlier had been drenched in the hot, dripping sun. The cold, quiet city glowed, blue-grey in the moonlight. Gone was the chitter-chatter of tourists and the pseudo-military guards keeping fascinating artefacts and tormented spirits behind well-bolted doors. Now the real residents were in charge.

Face after face on the tombs – the carved alter egos of the dead, their unblinking eyes watching us stumble; finally able to make their presence felt after the souvenir-hunters, ice creams and selfie sticks had been spat out. As we sped past people's former homes and possessions, down the now deserted paths of daily lives, under untrimmed trees and bushes, bats flew in front of our faces, swooping and screaming: 'Go back, go back.'

In the shadows, the map faded from sight. It had shown us what the guided tours hadn't; that there were miles of streets no-one was looking at; whole houses and temples not deemed worthy of a pithy description; palatial mansions almost completely intact; an amphitheatre empty, too far away for most people to walk to – or just not as quite big as the Colosseum. And then there was the cemetery.

Nobody comes to Pompeii to see an actual cemetery. The dead people they are interested in aren't buried in the ground; they are covered in plaster, their faces curled into horrific grimaces, illuminated by the continual flash of cameras on their glass-box prisons. Bodies destined to be forever incarcerated, highlighted on the map with a big red star. Maybe one day you'll go and stare at them. But perhaps afterwards you'll visit the real Pompeii, like we did.

Crackling stone; the stretched fingers of back streets; gardens once played in, now quiet and still; small pots where food was served hot and spicy, now cold and empty, the paraphernalia of daily life made special by the way its owners died. The bats, the rats and the bugs are the only life here now, along with us – and whatever else lurks around the corner.

Eventually: a small metal turnstile. And a man. A man! Apologies. We didn't realise the time… We thought we'd got locked in… We didn't know where to go… He shrugs: 'This is Italy. You go where you like.' Apparently it happens a lot; people getting lost. They – we – are all the same. None of us can read a map. But at least he was there to save us. 'A pleasant surprise, yes?'

And yet, as we leave and go back to the train, to a world of tourists, tickets and timetables, we can't help but be disappointed. Holidaymakers chat loudly, while locals chat louder. We miss Pompeii and the cold silky quiet of those who once walked its streets. Next time, we must try and stay longer. Next time, we must stay all night.

This haunting story was shortlisted in the Bradt Travel-writing Competition in 2015. We have since lost contact with the author.

TOUCHING HISTORY

Jenny Scott

Highly Commended 2016 'A Brief Encounter'

We landed in Athens in the early evening. In the airport bus through the city we ran into a Communist rally. Huge red flags waving and shouting and marching. Fresh from South Africa, with its terror of the 'communist threat', I felt a fearful shiver – a reality I did not understand. Cold to our bones, we took buses and creaking ferries as far south as we could, searching for warmth, and finally found ourselves in Agia Galini, on the southern shores of Crete.

No-one spoke a word of English in the seventies, except for things like '*ten minooot*' when you handed over a small fortune for a hot shower, or '*CLOSSHHED!*' every time you tried to get into a museum. We settled into our rustic little rooms at the boarding house overlooking the sea run by Maria and Evangelo, who happily pressed welcoming glasses of *ouzo* into our hands. We took to flying kites on the clifftops and long walks on the chilly beaches followed by *fasolatha*, a warming bean soup and *retsina* in the local *taverna*. Once on a cliff I stumbled on a gun emplacement in the bleak sunshine. Surrounded by lush grass and tiny spring flowers, cold forbidding steps led underground. And on the top step, chilling echoes of an ancient history. A soldier had scratched his name, 'Hans Werner' and the date, '13/2/42' into the wet cement.

But we couldn't stay forever, and bade farewell to Agia Galini. Only one bus a day, we headed for Rethymnon to continue our adventure. The journey would take two or three hours, climbing over the rocky mountains that bisect Crete, stopping at the charming white hill villages I could see as we slowly wound and climbed our way. Gentle rain streaked the dusty windows as fat old women with chickens clucking in baskets, old men with strange hats, and a few tourists filled the bus up.

Nearing the highest part of the mountain pass we stopped at Spili, and an old man climbed aboard. Dressed in long Cretan leather boots and breeches, as short as he was, his presence was compelling. He strode purposefully through the bus, loud and insistent, demanding of every pale face.

'*Sprechen Sie Deutsch?!*' '*Sprechen Sie Deutsch?!!*' He looked hard into my face and waited for my reply – there was no avoiding his dark, intimidating eyes. Again and again, through the bus, he commanded attention, until a slim blonde young man said yes, he '*Sprechened Sie Deutsch*'.

He was brought to the centre of the aisle exactly where we English were sitting. The old man made himself comfortable sitting sideways in front of us on a suddenly vacant seat. His gnarled hands rested imperiously on the heavy walking stick between his knees and he proceeded with his instructions.

'*Übersetzen!*' Translate! and settled back comfortably.

And then it started. Sentence by sentence, the story he intended us to hear unfolded. First in loud, insistent German, followed by our reluctant translator into English.

'*Ich war fünfzehn Jare alt.*' I was 15 years old.

'*Mein Vater und ich hielten Tauben.*' My father and I kept pigeons at the back of our house.

'*Die Deutschen kamen in unser Dorf.*' One day in 1943, the Germans came to our village. And so, we heard his story from a young German tourist, with his rimless glasses, his ashen face, his voice getting lower and lower, as he was forced into this public show. Still, we could make out the words as we all bumped along the little mountain road.

'Die Soldaten toteten alle meine Tauben.' First the soldiers, they shot all of my pigeons. Then all the men were taken up to the hill at the back of the village. There the men were forced to watch as the German soldiers herded the women and young children out of their homes.

We could barely hear any more, but between the barking German and the quiet faltering English, we could make it all out.

'Ich sprach perfekt Deutsch.' I spoke perfect German, so I survived. I watched as they shot all the men in the village.

'Darunter mein Vater.' Including my father.

'Brand im Dorf.' And then they torched the whole village.

The denim-clad German traveller didn't raise his eyes. It was impossible to look at him, and impossible not to listen, riveted by the first-hand account. The Germans followed exactly the same method each time there was a reprisal: massacred the men, obliterated the village, then allowed the women and children to return. The story ended as the we pulled into Rethymnon, the sun lowering over the harbour.

Eventually their village was rebuilt, and our storyteller – *'Ich habe immer noch Tauben'* – still kept pigeons.

And told his story every day on that slow and mournful bus.

***Jenny Scott** grew up on the tropical east coast of Africa. After motorcycle touring on that fascinating, if dark, continent she settled in the UK. Her writing started with a family/cookery blog which led to an MA in Creative Writing. A trip to Tikal inspired a love of ancient sites, and her travel list grows thousands of miles longer every day.*

585 BURY ROAD

Jennifer Thompson

Finalist 2021 'I'd Love to Go Back'

'Wakey, wakey, Bin Head!'

It was Mum or Dad that said it? I couldn't be sure. Jolted from dreams, I looked on to a rain-slicked road. Covid confinement had taken its toll, and I had fallen asleep against the living room window. Bin Head. Something about the old nickname reminded me of our trips to see family in Bolton. In any normal year, whatever normal is, we would have seen them at least twice by now. Cheek pressed against the cold glass, I imagined our car journey northwards.

I've never been good at reading maps, so the route between our southern home and Bolton is mapped by landmarks of my own designation. The perennial plains of the New Forest begin by the pencil-thin house mounting the dual carriageway. When kites replace buzzards circling the sky, I know Oxford is near. Modern megaliths exorcise boughs of cloud at Birmingham, where chimneys dominate the horizon. Beyond that is sleep shuttered. As the car crawls through city traffic towards Greater Manchester, an internal compass wakes me. Nearly there. Unfamiliar landmarks no longer blur beyond the boundary. Now, sites of interest pinpointed in my mind's map come into focus.

Groups of men gathered by a bus stop. The elders chatter, occasionally pointing up and down the road. Behind them, abandoned brick buildings have been converted into places of worship. Towering above them all, the gold dome of a new build. Mosque Road. What its real name is I do not know, but it is used by my family to navigate the addresses of cousins and great-aunts.

Lighting a cigarette, someone rolls down the passenger window near St Mary's. The peal of bells and sickly smell of snowdrops and bluebells. Against the dark brick of the church, I see a small girl in purple satin. Myself as a child, hitching my skirt to avoid the ants as my cousin, the bride, sneaks a cigarette of her own behind a bouquet of sweet peas.

The pub where Dad and the uncles gather to play boules rounds into view. White walled, black beamed. As the men down pints and mock each other's play, cribbage is dealt by the aunties. My mouth waters. Someone will have bought pasties and pies from the baker. Wrapped in a white carrier bag, they emerge golden and piping. My deepest secret? I prefer these pasties to the ones we get down south.

Next stop, Bury Road. Lifting my head from the backseat window, I look beyond the glass. We park outside a row of identical terraced houses. Each flat-fronted and featureless, except for one protuberant window and an awning that spreads over the doors of the adjoining house. Separated by vast red brick, the windows of the upper floors are as close to belonging to their neighbour's home as their own. Looking at them from the road, each wears a wide-mouthed grimace.

In the narrow doorway of 585, apron folded down to her skirt, my grandmother stands. Her grey hair gleams at the edges and, dressed in pink, grey and powered blue, she shines against the dull russet of the smog-stained brickwork.

After family, it is the bricks that fill my head when I think of Bolton. It is a red I cannot find anywhere else. Not in winter gloaming, before the sun melts into the horizon. Nor can I find it in the hue of a ripening bruise or the speckle of drying blood. Mills, churches, schools. All built by bricks now housing hierarchies. They laid themselves down as foundations so manmade mountains could court the clouds. The Empire State and Blackpool Tower.

My mother's head turns towards number 585 from the passenger seat. She repeats the story of a Christmas years ago when, ignoring my gifts, I fashioned a hat from a metal bin. Trying to remember every detail, I can only conjure the carpet and dark stairs. Grandma isn't there anymore, you see, but remains the most vivid landmark in my memory.

She smelt of the warmth at the back of an airing cupboard, and of time-worn second-hand books. Her hands were soft. When I miss her, I stretch cling film over my own to mimic their texture. The skin on the back of them was crêpe-paper thin, and the blue veins beneath spread into her elegant fingers like the roads on Dad's AA map.

'Wakey, wakey, Bin Head!'

Cheeks pressed against the glass from within the walls of lockdown, I imagined 585 Bury Road. Grandma isn't there anymore, but we'll drive along that red-bricked road again, and I'll see her in the doorway once more.

Jennifer Thompson *lives in southwest England. She holds an MA in Travel and Nature Writing, was shortlisted for the Porthleven Prize and, in 2021, became the inaugural Emerging Writer in Residence with the Causley Trust. Her work has been published in print and online, including* The Pilgrim, Guernsey Press, Oh, Nature Makes Us Better *and* Adventurous Ink. *She is a regular contributor to the* Marine Biologist.

8
PASSPORTS & PRIZES

"My prize was a trip was to Uganda – the highlight was scrambling through the forested hills of the Bwindi Impenetrable National Park to see gorillas."

Dom Tulett

Historically, the prizes in the Bradt competition have been spectacular, with trips and tours and special itineraries to extraordinary places. Prize winners also win a commission to write a piece about their trip and these have been published in various outlets including the *Independent on Sunday*, *Timeless Travels* and Wexas *Traveller* magazine.

Runners-up have been offered places on travel-writing courses such as Travellers' Tales writing weekends in northern France, and Bradt's own annual travel-writing seminar.

I wrote about the desert for the 2019 competition and won a trip to the Arctic! My prize trip was to northern Finland – to the land of the wolf and the bear; fiery skies, deep cold and ancient, nomadic ways. The trip included a moonlit sledge ride, snowshoeing and a snowmobile safari.

Here are some snippets from other winners' experiences.

Celia Dillow

In 2014, **Elizabeth Gowing** won a trip to **Nova Scotia, Canada**. Her stand-out memory is a watery one. A frightening experience in a canoe as a child left her wary of the free canoes outside her prize accommodation. But eventually the glassy stillness was too tempting and she quietly untied one of the canoes and launched off into a feeling of great pride and calm – and closure.

Dom Tulett won a trip to **Uganda** in 2016: nights by the shore of Lake Victoria in Entebbe; game drives; a dawn hot-air balloon ride in the Queen Elizabeth National Park and a boat safari on the Kazinga Channel. His story of the gorillas in the Bwindi Impenetrable National Park won the *National Geographic* Travel Writing Competition the following year.

In 2013, **Cal Flyn** won a delicious gastronomic tour to Istria in **Croatia**. She says they ate so much rich food in the space of three days (ten-course tasting menus; truffles they'd watched being sniffed out by dogs – freshly shaved over scrambled eggs; wine before noon; and shots of olive oil for tasting) that they were left gasping and couldn't eat for more than 36 hours afterwards!

Liam Hodgkinson won a trip to **Iceland** in 2015. He says it was desolate and beautiful. They drove through the lava fields into a storm and the road was so hidden by the rain that they felt like they were flying, not driving.

In 2005, **Sara Evans** won a long weekend in wintery **Slovenia**. Her stand-out memory was a torch-lit boat tour of the ethereal and extraordinarily biodiverse emerald-green subterranean lakes of the Križna Cave, in the south of the country.

Marie Kreft's prize, in 2010, was a week's holiday in **Gozo and Malta**. Her most memorable day was a visit to the megalithic temple complex of Hagar Qim and Mnajdra on Malta's west coast, which she wrote about for her commissioned piece in the *Independent on Sunday*. Because she was writing about the temples, a local guide was pleased to add much greater depth to her understanding of the mysterious site.

In her competition entry, she had written about being engaged to Steve. In a sweet little coda, the story happened to be published on the morning after their wedding. Steve tipped off both families and they all ordered the newspaper to their hotel rooms. She came down to breakfast to find everyone reading it!

Louise Heal was sent to **Sri Lanka** in 2007, when there were few tourists in the country. On Remembrance Sunday, she went out before breakfast with her poppy pinned to her bag. She took a rickshaw to Sigirya, the 5th-century rock fortress, and walked through empty water gardens with a guide. They climbed the rock in silent solitude, ducking under hornets' nests and admiring frescoes on the way. Standing alone on top of the fortress, her guide took a picture of her which still stirs powerful memories.

Alan Packer's prize was a trip to **Lapland**, to the Ice Hotel. One highlight was to hear the new organ in the ancient church of Jukkasjärvi. Sami artist Lars Levi Sunna decorated the organ and keyboard, inlaying reindeer horn and birch wood on a pine background. The range of organ tones was extended to celebrate Sami life: the Sami drum, forest birds and the twinkling of winter stars. The church guide let Alan's wife, Mary, play the organ – one of those unexpected thrilling moments that occur when you travel.

Destinations for prize winners include the following, courtesy of various partners including the Croatian National Tourist Office, the Istria Tourist Board, Travellers' Tales, the Canadian Tourism Commission, Pas-de-Calais Tourism, Discover the World, TravelLocal and Wexas Travel.

2004 Mostar, Bosnia & Herzegovina

2005 Lake Bohinj & Ljubljana, Slovenia

2006 Budapest & Lake Balaton, Hungary

2007 Sri Lanka

2008 Kyrgyzstan

2010 Gozo & Malta

2013 Istria, Croatia

2014 Nova Scotia, Canada

2015 Fly-drive in Iceland

2016 Lion & gorilla safari in Uganda

2017 Ice Hotel in Swedish Lapland

2019 Inari, Finland

THE BRADT STORY

In the beginning

It all began in 1974 on an Amazon river barge. During an 18-month trip through South America, two adventurous young backpackers – Hilary Bradt and her then husband, George – decided to write about the hiking trails they had discovered through the Andes. *Backpacking Along Ancient Ways in Peru and Bolivia* included the very first descriptions of the Inca Trail. It was the start of a colourful journey to becoming one of the best-loved travel publishers in the world; you can read the full story on our website (www.bradtguides.com/ourstory).

Getting there first

Hilary quickly gained a reputation for being a true travel pioneer, and in the 1980s she started to focus on guides to places overlooked by other publishers. The Bradt Guides list became a roll call of guidebook 'firsts'. We published the first guide to Madagascar, followed by Mauritius, Czechoslovakia and Vietnam. The 1990s saw the beginning of our extensive coverage of Africa: Tanzania, Uganda, South Africa, and Eritrea. Later, post-conflict guides became a feature: Rwanda, Mozambique, Angola, Sierra Leone, Bosnia and Kosovo.

Comprehensive – and with a conscience

Today, we are the world's largest independently owned travel publisher, with more than 200 titles, from full-count and wildlife guides to Slow Travel guides like this one. However, our ethos remains unchanged. Hilary is still keenly involved, and we still get there first: two-thirds of Bradt guides have no direct competition.

But we don't just get there first. Our guides are also known for being more comprehensive than any other series. We avoid templates and tick-lists. Each guide is a one-of-a-kind expression of ar expert author's interests, knowledge and enthusiasm for telling it how it really is.

And a commitment to wildlife, conservation and respect for local communities has always been at the heart of our books. Bradt Guides was championing sustainable travel before ar other guidebook publisher.

Thank you!

We can only do what we do because of the support of readers like you – people who value less-obvious experiences, less-visited places and a more thoughtfu approach to travel. Those who, like us, ta travel seriously.